Relativity in Our Time

To the Memory of My Father
Samuel Sachs

Relativity in Our Time

From Physics to Human Relations

Mendel Sachs
Department of Physics and Astronomy
State University of New York at Buffalo

Taylor & Francis
London • Washington, DC

UK   Taylor & Francis Ltd, 4 John St, London WC1N 2ET

USA   Taylor & Francis Inc., 1900 Frost Road, Suite 101, Bristol, PA 19007

British Library Cataloguing in Publication Data

A catalogue record for this book is available from the British Library

ISBN   0-7484-0117-2 (cased)
       0-7484-0118-0 (paper)

Library of Congress Cataloging-in-Publication Data are available

Cover design by John Leath, illustration by Daniel E. Sachs

Printed in Great Britain by Burgess Science Press, Basingstoke, on paper which has a specified pH value on final paper manufacture of not less than 7.5 and therefore 'acid free'.

Contents

perception of space versus its abstract characteristic. 'Space' in the
language of science.

concept of a universal force field. *Dialogue*: On logical and aesthetic
bases for the unified force field concept. Comparison of the
evolution of ideas in science and the evolution of the human society
– from the model of 'thingness' to that of 'holism'.

Preface

One of the most significant advances in twentieth century physics, indeed in the entire history of science, was Albert Einstein's discovery of the theory of relativity. Though its mathematical implications have been necessary in much of modern physics, there are some widespread misconceptions about the meaning of this theory. In this monograph I explain my view of the fundamental meaning of this scientific theory – the meaning which I believe Einstein intended, according to his own writings, especially after it had matured to the stage of general relativity[1].

In addition to the implications of the theory of relativity with regard to the 'inanimate' features of the universe – the science of physics – the underlying philosophy, when followed to its logical extreme, seems to me to carry over to the field of human relations. For it is a philosophy that implies a unification of man with nature, humanism, and mutual respect between all components of the world.

I will attempt to convince the reader of this conclusion throughout the text, and thus as we proceed with the exposition of ideas in this book, I will relate this philosophy, not only to the applications to the physical world of inanimate objects, but also to implications with regard to human relations – as seen through the eyes of a physical scientist, rather than from the views of a professional sociologist or psychologist.

It would be presumptuous of me to allude to any real expertise in the various fields of human relations. Still, I do feel that as a member of the human race it might be of benefit to our understanding of the world if the physical scientist spent a part of his or her thinking time on problems that underlie the social sciences, just as I feel that it could benefit the social scientist's quest for further understanding with regard to human relations, to look (just as non-professionally!) at some of the ideas that underlie the physical sciences. These reflections are based, in part, on my belief in the Spinozist view of the oneness of man with nature – in fundamental terms – implying that the more we can unify the theories about man's behaviour in society with those we have discovered concerned with the physical universe, the closer will we be to a valid understanding of the real world[2].

This is not to say that thus far our search for fundamental truths about the natural world has not had an important practical impact on our way of life, our attitudes and well-being. Indeed, the slightest glimpse at the history of the human race reveals that great practical advances have been made: in medicine, to prolong life with years of better health and comfort; in technology, to provide us with a more affluent life, more leisure and, unfortunately, a more efficient means of self-destruction!

But aside from practical by-products, the scientist trying to understand the physical world, for the sake of understanding itself, would maintain that the philosophical insights gained by humankind have raised our cultural sights; they have opened the door to increased understanding of ourselves, as inseparable components of the world, hopefully toward a life of peace, mutual respect and oneness.

Many express doubts that the advantages I claim really do follow from purely intellectual endeavours, and ask: What is 'pure knowledge' really good for? To answer that we seek a basic understanding of the world because history teaches us that such studies have always (eventually) led to practical applications, contradicts my initial assertion that this is a pursuit of understanding for the sake of understanding itself! Indeed, with the motivation of practical applications in mind – no matter how far in the future they may come – science, as any other pursuit of ideas, would slowly become corrupted of its original purpose, until it would be dead! This can be likened to the gradual petrification of a tree – it may look like a living tree after the process has been completed, but in fact it would then only be inorganic, dead stone!

I believe that the answer to the question is in part subjective. My answer is that society should support the activities of purely intellectual pursuit, because it is natural for the human being to explore his or her curiosity, and in my view, this has positive value. The act of killing another human being may also be a natural function, but in my view it has negative value! My criteria for distinguishing positive from negative values are that positive actions relate to: the well-being of society, of which the individual is an integral, inseparable component, rather than a dispensable 'part'; to humanism; and to the holistic concept of the oneness of all of nature.

Further implications follow from these criteria. One is the positive value of the freedom of the individual – to the extent that an individual's actions neither harm nor restrict the freedom of others. Another is the rejection of the (in my view) negative value of the policy that 'the end justifies the means'. I do not look at the world, fundamentally, in terms of a time-evolving entity. I see it, rather, as a basic existent, of which humankind is a particular manifestation, with the feature of 'influencing' and 'being influenced' at one stroke. With this view, then, all our actions that would be positive must be for the well-being of the world, as it exists. Actions that lead to the removal of freedom and inhumane treatments of fellow members of the human society, as well as self-motivated destruction of our natural environment, cannot be justified at any time, no matter what ends are claimed! If we are to learn anything from the history of our society, the first lesson is that a policy based on this notion that the 'end justifies the means' can only lead to failure in reaching the ultimate (sometimes well-intentioned) goals. This route has led mostly to the attainment of totalitarian control in the hands of a few leaders, exploiting their populace in order to maintain power, while paying lip-service to altruistic principles[3]!

The existential view of the world, that I take, is not that it is a physical system in space and time. Rather, it is based most fundamentally on an underlying order, in terms of the mutuality of its (inseparable) components. This view implies, to me, that every action in every day of one's life must be considered from the standpoint of the entire society. I do not claim that such a Utopian society can be achieved, but I do believe that it would be progressive to strive toward the attainment of attitudes in this direction.

One of the important developments of 20th century science that teaches this philosophy is the theory of relativity, when expressed in its full form. For, with the approach of this theory, we have the implication that the world is continuously one – a closed system of inseparable components. However, to convince our fellow human beings of the truth of this assertion, it is necessary to do more than philosophize. We must exploit this philosophy in a precise way in order to demonstrate the truth of its assertions. If this can be done with regard to the 'inanimate' aspects of the world – the apparent things that we see and experience about us – from the cosmological domain of the universe as a whole to the microscopic domain of elementary particle physics, then we would have at least made a start toward extending these principles to the domain of human relations.

The main discussion of this book is based on a seminar course, designed for undergraduates from the sciences *and* the humanities, on the underlying philosophy of Einstein's relativity theory. The inclusion of students from the humanities always enlivened the atmosphere with the sorts of questions and comments that science students do not generally raise, regarding some of the implications of this philosophy in societal problems. My answers to their questions, as well as the comments of the other participants in the seminars, were not those of professionals in sociology, anthropology or psychology. Rather, they were the natural responses of lay people or students in these fields, who are generally interested in the implications of the pursuit of ideas that may eventually be applicable to understanding ourselves more completely and in satisfying our natural curiosity about the world in general. I have incorporated some of this discussion in the body of the text, but most is in the form of Question–Reply sections towards the end of the chapters.

When each set of seminars was completed, I felt that most participants indeed came away with some basic understanding of the ideas that underlie the theory of relativity – not its mathematical structure which, of course, is necessary for the professional physicist to exploit technically the theory further – but rather the logical connection of ideas that underlie this very great discovery in contemporary science. I believe (perhaps wishfully!) that most participants were convinced that pure science, for its own sake, is indeed relevant to society, primarily with regard to a broadening of our culture. This activity, in my view, leads to an enrichment of human values and helps to reveal to individuals their true authenticity, through the climate of freedom that it creates. My main aim in writing this book is for the same sort of enthusiasm to carry over to the reader.

Mendel Sachs
Buffalo, New York

NOTES

1 A clear statement of Einstein's interpretation of the theory of relativity, that he had settled on as his theory matured, is given in his *Autobiographical Notes*, in *Albert Einstein – Philosopher–Scientist* (edited by P. A. Schilpp) (Open Court, 1949).
2 Spinoza's main philosophical views were expressed in: B. Spinoza, *Ethics*, R. H. M. Elwes, transl. (Dover, 1955). Also see: H. A. Wolfson, *The Philosophy of Spinoza* (Schocken, 1969).
3 A well-known contemporary political philosophy that maintains the policy that 'the end justifies the means' is Communism. It is indeed apropos that in our own time, in 1991, one of the most powerful nations in the world, whose government professed to be based on this political philosophy – the Soviet Union – has abandoned it, renouncing Communism and its totalitarian approach. This was not only because of its lack of humanism, but also for the pragmatic reason that it did not work! It took 74 years for the Communist form of government of the USSR to break down – as an ultimate consequence of its ideology.

1

Introduction

Before coming to the main text of this monograph, I would like to make three preliminary remarks about relativity theory. First, in the context of the history of science, Einstein's theory of relativity did not appear at a particular point in time, disconnected from all preceding developments of ideas in physics. Indeed, this theory has some very important roots in classical physics and in the ideas of the ancient times.

Some of the most important precursors for Einstein's theory of relativity are:

- Galileo's principle of inertia[1];
- Galileo's concept of the subjectivity of all motion and spatial measure to underlie the objectivity of a law of moving matter, called, 'Galileo's principle of relativity'[2];
- Newton's third law of motion, implying the elementarity of a closed system of matter and the relativity of the reference frame of one component of matter in mutual interaction with another, in accordance with the statement of his law: for every force exerted by one body of matter on another, there is an equal and oppositely directed force exerted by the latter body on the former; and
- important precursors for the theory of relativity in ancient Greece i.e. concepts of invariance that one may see (in abstract form) in the metaphysical views of Plato – his concept of 'forms' – and the holistic view of Parmenides, as well as the elementarity of process in Heraclitus[3].

Second, while the mathematical expressions can become quite complicated, especially in its most general form (in general relativity), this theory is extremely simple from the conceptual point of view. It is the purpose of this monograph to discuss the conceptual and logical structure of the theory, rather than its mathematical language. For to comprehend its conceptual content is to understand the theory of relativity. It was Einstein's contention, as I will attempt to demonstrate below, that because of its extreme conceptual simplicity, the theory of relativity should be comprehensible to not only the students and professionals in physics, but also to any lay readers who wish to broaden their comprehension of the natural world.

Therein lies a major confusion among contemporary scientists, as well as the lay public – a confusion between 'mathematical simplicity', on the one hand, and 'conceptual simplicity' on the other. That the theory of relativity is not mathematically simple, in all of its ramifications, does not imply that it is equally unsimple from its conceptual side!

Indeed, one of Einstein's primary reasons for his strong faith in the objective truth of the theory of relativity is its logical simplicity (not its mathematical simplicity!). In a letter that Einstein wrote to Louis de Broglie in 1954, he said[4]:

> Die gravitationsgleichungen waren nur auffindbar auf Grund eines rein formalen Prinzips (allgemeine Kovarianz), d.h. auf Grund des Vetrauens auf die denkbar in grösste logische Einfachheit der Naturgesetze.

> [The equations of gravitation were able to be discovered only on the basis of a purely formal principle (general covariance) that is to say on the basis of the conviction that the laws of nature have the greatest imaginable logical simplicity.]

Third, Einstein's meaning of relativity theory evolved from his earlier conception, 'special relativity', to an entirely different conceptual view, when 'general relativity' appeared. His conception of special relativity was based on an epistemological stand of operationalism and positivism – asserting that the only meaningful statements about the world must be directly tied to empirical or operational responses that human beings may have – and where the model of matter is in terms of atomism. His later epistemological stand in general relativity is based on the view of realism – wherein the assumption is made that there is a real world, independent of whether or not there are human beings around to perceive its features – and where the model of matter is most fundamentally in terms of continuity rather than atomism. In this view, it is indeed possible to draw conclusions about the real world that are not directly tied to empirical evidences, though these statements may lead, in a logical chain of conclusions, to empirical predictions[5].

THE BASIC PREMISE OF RELATIVITY – THE PRINCIPLE OF COVARIANCE

The basic idea of the theory of relativity that convinced Einstein of its extreme simplicity is his principle of covariance, also referred to as the principle of relativity.

This principle asserts that the laws of nature must have expressions independent of the frame of reference in which they are represented – from any particular observer's view. This is equivalent to saying that the laws of nature are totally objective.

We see, then, that the theory of relativity is based on a premise that is a law about laws, rather than a law that deals directly with physical phenomena. The idea about the objectivity of the laws of nature is, however, not really that new! For how could a law be a law, by definition of the word 'law', if it were not totally objective?

Thus it seems at this stage of the discussion that the theory of relativity is based on a premise that is tautological – such as the statement that 'woman is female'! If this were indeed the case, then Einstein's theory of relativity would not be a scientific theory, for a scientific truth must be contingent on nature, whereas a tautological truth is a 'necessary truth' – it cannot be anything other than what it asserts.

Nevertheless, the principle of covariance of relativity theory is not really a tautology because it depends on two tacit assumptions that are indeed contingent on nature:

- there exist universal scientific laws that prescribe the logical connections between causes and effects; and
- these scientific laws may be comprehended by us and expressed in a precise way – sufficiently so as to allow us to test their conclusions in experimentation as well as testing their logical consistency.

The first of these tacit assumptions is based on an idea sometimes referred to as 'the principle of total causality' – the idea that for every physical effect in the world there is a logically connected physical cause. Of course, this assertion is not necessarily true. However, it seems to me that it is the credo of the scientists, since it is their very purpose, as scientists, to search for the causal connections of the physical manifestations of the universe, in any of its domains, in order, in turn, to gain in our comprehension of the fundamental nature of matter.

The second tacit assumption – the idea that we can comprehend and express the physical laws – is also not necessarily true. Perhaps it is arrogant for us, the human beings on this planet Earth, here in this rather insignificant corner of the universe, to make such a bold claim. Nevertheless, I believe that the history of science attests to the fact that indeed we have made some progress in the direction of increased understanding of the infinite universe, over the past millennia. Thus we have some confidence that it is possible for us to attain increased understanding of the objective physical laws of the universe – miniscule as our accumulated understanding may be at any particular stage of our history (including the present!), compared with all that there is to understand. This assertion is along the same line of thought as Einstein's, when he said: 'The most incomprehensible thing to me is that we can comprehend anything about the universe!' Thus he had confidence that we can indeed comprehend something of objective reality, small as this understanding may be compared with a total understanding – which we could never achieve, because of our finiteness.

Nevertheless, it is our obligation, as scientists, to continue with the task of gaining in our understanding of the real world.

THE ROLE OF SPACE AND TIME IN RELATIVITY THEORY

The assumption that we can express the laws of nature, means that we can find an appropriate language for a precise way to represent the laws. This idea has led to the introduction of the space–time in the theory that is different than its use in the earlier physical theories. For in the classical views, space and time are there, whether or not matter is present. One then 'puts matter into space and time', as one might put socks into an empty drawer. But in the view of relativity theory, space and time are not 'things-in-themselves' – rather they are only related to a language that is invented for the purpose of facilitating an expression of the laws of matter.

The most convenient language that we have discovered so far in the history of physics, to represent the laws of motion, is in terms of four continuously variable parameters, playing the role of the 'words' of this language, and a logic that connects them (analogous to the syntax of verbal language, such as the 'subject–predicate' relation), to give meaning to the expressions of this language. The logic of the space–time language of relativity theory is in two parts: algebra and geometry. [At the present stage of Mathematics, mathematicians believe that all theorems of algebra and geometry merge into a common set of theorems, and thus their actual separation is artificial. However, for the purposes of this exposition it will be convenient to consider them separately.]

The principle of covariance asserts that the laws of nature must be totally objective – meaning that their forms must be independent of the space–time reference frame in which they are expressed, from any particular observer's point of view. The space–time language itself is relative to the reference frame in which it is expressed – hence the name of this theory! Still, the primary focus of the principle of covariance is on something that is absolute rather than relative – this is the invariant law of nature itself.

In the early stages of this theory, its name led to the erroneous impression that this approach in science is based on the philosophic view of 'relativism' – the idea that all knowledge is relative only to the 'knower' – i.e., that there is no objective knowledge to talk about. Of course, Einstein never had this view in mind – his approach was just the opposite, where one focuses on the invariant (objective) law of nature. To avoid the confusion, Einstein tried to rename his theory 'invariententheorie' (theory of invariants), implying a focus of this theory on absoluteness rather than relativeness. However, he eventually rejected the name change because of further confusion he thought it might entail[6].

The role of space and time in relativity theory is then to serve as a language whose sole purpose is to facilitate a subjective expression (i.e.

relative to the reference frame) for objective laws of nature (i.e. laws that are in one-to-one correspondence in regard to their expressions in all possible frames of reference.

To exemplify the notion that not everything is relative, we note that one thing that cannot be relative is the universe itself. For the universe, by definition, is all that there is, therefore there is nothing else to be relative to! This absoluteness of the universe, in turn, entails the absoluteness of its basic characteristics, as expressed in the laws of nature. Thus we see that, logically, the absoluteness of the universe as a whole implies that the laws of nature must have forms that are absolute – that they are 'covariant' with respect to changes to any reference frame – that is, any scientific investigator would see that the law for any particular phenomenon in all possible reference frames relative to his own must be in exact correspondence. But this is just Einstein's starting premise of the theory of relativity – the principle of covariance.

This idea is entirely analogous to the subjectivity of verbal languages in order to express objective meanings. For example, before Newton's discovery of the law of universal gravitation, the English may have expressed the law of gravity as: 'Whatever goes up, must come down'. The French, to express the same meaning, would have said: 'ce qui s'élève doit descendre'. The languages are different, but they express precisely the same idea.

Einstein's principle of covariance then asserts that if a particular set of relations is indeed a universal law of nature, it must be independent of the reference frame in which it is expressed. If, in the preceding example, the English would have said something slightly different, such as: 'Whatever goes up usually comes down', but the French said the same thing as before, the principle of covariance would be in violation, implying that the scientists from both 'frames of reference' should investigate further until they could come up with an objective statement about 'things that go up'.

The language translations between English and French in this example, applied to all other possible statements, so as to preserve the meanings of the sentences (in one-to-one correspondence) then forms a 'transformation group' – in the language of algebra.

It is interesting to note that the full set of such 'translations' of the mathematical space–time language in physics is more precise than the translations of verbal languages to preserve the meanings of sentences, e.g. between English and French. This is because the translations of verbal languages entail meanings expressed with words and body gestures of one language that are tied to a particular culture that is not easily translatable into ideas in terms of words and body gestures of the language of a different culture. But there is no such difficulty encountered in the translation of mathematical languages because of their increased precision, and because they are all based on a single (scientific) culture. Even so, the mathematical languages are not as rich as verbal languages, certainly not at our stage of development of the human culture. [I believe that they will never match, because human feelings will probably never be expressible in mathematical terms!]

Implications of the space–time logic in the physical laws

The universality of the speed of light

A logical implication of the principle of covariance is that purely spatial and purely temporal measures are not objective entities in themselves. Indeed, it was discovered by Einstein (as we will discuss more explicitly in the next chapter) that in order to express a law of nature covariantly, a spatial measure in one frame of reference must be expressed as a particular combination of spatial and temporal measures in other frames of reference if the law that they express is to be in correspondence in the different reference frames. Similarly, a purely temporal measure in one particular reference frame must be expressed as a mixture of spatial and temporal measures in other (relatively moving) reference frames. Thus, it is necessary in this theory to express the words of the language as 'space–time' measures, rather than space measures separately from time measures.

Now if one must generally express space and time measures in this fused way in the laws of nature, then they must be defined with the same units; instead of referring to time measures in the different reference frames as $(t, t', t'', ...)$ seconds, we should refer to them as $(ct, ct', ct'', ...)$ centimetres, where c is a universal conversion factor, independent of reference frame, with the dimension of centimetre per second – a universal speed. If the conversion factor c would not be independent of reference frames, then it would not be possible to separate physical extension from duration, that correlate with the space and time measures in the predictions of a theory. Such a separation is, of course, an empirical requirement of the theory. Thus it follows that the speed c is a constant, independent of the reference frame in which it is expressed.

At this stage of the argument, there is nothing to tell us the magnitude of this constant speed, c. It was when Einstein studied the laws of electromagnetism, in the form of Maxwell's equations, that he identified this constant with the precise value of the speed of light in a vacuum.

Since c is a universal constant in the expression of the space–time language for all physical laws, not just electromagnetism, it must be the same conversion factor in the expression of all other laws of nature, when they are expressed covariantly. That is to say, all laws of nature, according to the theory of relativity, are supposed to preserve their forms in transformations to all possible reference frames, just as the language translation between English and all other languages must preserve the meanings of all possible sentences.

Thus we have seen that the universality of the speed of light is a logical consequence of the principle of covariance, and not an independent axiom of the theory of relativity – leaving the principle of covariance as the only basic premise of the theory.

In his initial papers on this theory, Einstein said that the universality of the speed of light is an independent axiom of this theory, in addition to the principle of covariance. Showing that the latter is truly the only independent axiom of the theory then increases its simplicity, thereby satisfying further Einstein's criterion of maximum simplicity for the physical laws.

Continuity replaces atomism – an implication of the algebraic logic of relativity theory

The algebraic logic of the space–time language relates space–time points to each other in the sense of rules of combination, countability, compactness, and so on. The most precise way to express this logic is in terms of a 'symmetry group' and its irreducible representations. Because space–time is a continuous set of parameters, the symmetry group of relativity theory is a continuous group. A profound implication of this algebra is that the solutions of the laws of nature are continuous functions of the space–time variables. These are called the 'field variables' and are, most fundament:..ly, to represent matter in all possible domains – from the physics of elementary matter (protons, electrons, etc.) to the physics of the universe as a whole, the domain of 'cosmology'. The field variables are then the 'dependent variables', that solve the laws of nature (the field laws) while they are 'mapped' in the space of 'independent variables' (the space–time coordinate system). Such a theory is then based on the 'field concept'.

The field concept, logically implied by the principle of covariance of relativity theory, then leads to the idea that any domain of matter has no discrete spatial cut-off. It then follows that, most fundamentally, the world is not atomistic. What appear to be the elementary particles of matter, according to this view, are in reality spatially peaked modes of a continuum matter field – analogous to the ripples of a continuous pond. This is indeed a physical consequence of the algebraic logic of space–time in the theory of relativity, where atomism is replaced with continuity, everywhere[7].

Implications of the geometric logic of space–time in relativity theory – a curved space–time and the principle of equivalence

The geometric logic of the space–time language relates the space–time 'words' to each other in the sense of relative congruence, parallelism, mapping, and so on. It then follows that if this language is to reflect the continuous variability of the matter fields in space–time, then the relations of the space–time points to each other must be correspondingly variable. However, the relations between the points of a euclidean space are the same everywhere. Einstein then recognized that euclidean geometry is an inadequate logic for the language of the laws of matter in relativity theory.

The new logic that he adapted was Riemann's differential geometry. It has the feature that as the parameters that represent the presence of matter are continuously diminished, the riemannian geometry correspondingly approaches a euclidean geometry. This is an asymptotic approach, never in principle reaching the limit, so long as we are describing existing matter (i.e. the subject of physics, *per se*). With this view, the limiting euclidean geometrical system can only represent the language for the laws of a perfect vacuum, everywhere[8].

An interesting implication of this result is that the case of euclidean geometry, and the resulting formulas of 'special relativity theory' is an ideal, though unachievable limit of the 'theory of general relativity' – as we will discuss in more detail in Chapters 17 and 18 on general relativity. The latter (unachievable) limit physically corresponds to a vacuum, everywhere, i.e. where no matter exists, anywhere. Though this is not a real situation, it is indeed a valid mathematical approximation to assume this limit in many applications in physics, where the theory of special relativity provides useful formulae. Still, it is important to keep in mind that this is an approximation for something else, and we will discuss this in more detail in Chapters 17 and 18.

Another interesting feature of the riemannian geometrical space–time, setting it aside from a euclidean geometry, is that the paths of minimal (or maximal) separation between any of their points – the 'geodesics' of space–time – are continuously variable curves, rather than the straight lines of euclidean geometry. The complete family of curved geodesics is called a 'curved space–time'.

A physically significant aspect of the geodesic is that the free motion of a test body, unimpeded by an external force, must follow such a path. For the free body to move along any other path would require energy from the outside to change that path – i.e. in the latter case, the test body would not be 'free'. Thus, with the geometrical logic of Euclid, as in special relativity or in classical physics, the free body's natural motion would be along a straight line path, relative to any observer (who is inertial relative to this body, i.e. at rest or in motion with a constant speed in a straight line, relative to the test body). This is indeed an expression of Galileo's principle of inertia or, as expressed a generation later, Newton's first law of motion.

On the other hand, with the imposition of the riemannian (rather than euclidean) geometry for the space–time language of the laws of nature, the geodesic path for the 'free body' (that is, a test body unimpeded by external influences aside from the source of the curvature of the space–time) would be a curve rather than a straight line.

From the classical view, an observation of the curved path trajectory for the body would imply that an external force is causing the motion. We see, then, that there is a sort of equivalence between explaining the motion of a body on a curve as, on the one hand, due to an external force acting on it (as in classical physics) or, on the other hand, describing the motion of this body without direct reference to external forces, but rather as moving along the geodesic of a curved space–time. In my view, this is the most general expression of Einstein's principle of equivalence[9].

If the particular curvature of space–time, in representing the existence of a matter field, predicts the path of a test body – such as the orbital path of a planet of our solar system relative to the sun's position, or the path of a galaxy relative to another galaxy in the heavens, or the path of a rock falling from a cliff – in better agreement with the empirical facts than the classical predictions of the newtonian theory of universal gravitation, then this new theory would meet the criteria to supersede the older theory of gravitation. This goal was indeed that of the theory of general relativity in its first stages.

It then remains to determine the relation between the geometrical fields of space–time – as the 'words' of this language – and the effects of the material content of the system. This goal of Einstein's investigation led him to the 'Einstein field equations': the matter fields appear as sources in a particular form on the right-hand side; and the solutions of the equations that appear on the left-hand side then lead to the prediction of the geodesic path that represents the existence of the matter source of its curvature, giving, in turn, the path of the test body. In this way, Einstein's theory of general relativity met the criteria to unseat Newton's theory of universal gravitation, as a new and improved explanation of the gravitational manifestation of matter. For Einstein's theory not only predicted all of the correct results of the classical theory, it also predicted new effects in agreement with the data that were not even anticipated qualitatively by the classical theory. We will discuss these developments in greater detail in Chapter 19.

Implications of a unified field theory in general relativity

Though the initial stages of Einstein's field theory of general relativity successfully superseded Newton's theory, that had reigned for 300 years, it was not meant to explain only the phenomenon of gravitation. From Einstein's view, general relativity was to be taken as a general theory of matter, based, axiomatically, on the principle of covariance – an assertion that applies to all laws of matter, not just gravitation.

Now, the principle of covariance implies that all theories of matter are based on the continuous field concept, wherein the basic field variables of matter are continuous functions, mapped in a curved space–time. This approach to matter automatically rules out the model in terms of discrete atoms, though the modes of the continuous matter field can be sharply peaked, spatially, but not actually discrete anywhere. Also, 'special relativity' may be taken as an asymptotic approximation for general relativity – a limit that is approachable as closely as we please, but not in principle reachable.

There is also a logical implication in relativity theory that one type of force, characterized by a particular sort of force law, cannot suddenly turn off, discretely, while another type of force replaces it then and there, at the edge of a particular domain. To preserve the covariance of the theory, all possible force manifestations of matter must be entailed in a given matter field that, in turn, is a solution of the laws of matter, i.e. the field equations.

For example, it was discovered by Michael Faraday, in the 19th century, that the electric field of force and the magnetic field of force are indeed unified in the sense that that magnetic field is not more than an electric field in motion. But since motion, *per se*, is a purely subjective aspect in the description of interacting matter, (i.e. if Peter moves relative to Paul it is equally true that Paul moves relative to Peter – a fact pointed out, originally in modern physics, by Galileo, in his discovery of 'Galileo's principle of relativity') it was concluded that the electric field of force must also be a magnetic field of force in motion. Thus, the electric and the magnetic forces are really a unified 'electromagnetic field of force' appearing to be purely electric under special conditions (where the observer is at rest relative to the electrically charged matter). Similarly, the electromagnetic field of force appears to be purely magnetic under other conditions whereby the electric field of force is masked. It was then Faraday's idea that all possible force manifestations of matter should be aspects of a unified field of force, showing itself to be one sort of force or another under correspondingly different types of conditions of observation, yet objectively deriving from a single unified field[10].

Einstein was the primary investigator to continue Faraday's 19th century investigations of the unified field theory into the 20th century – since the same idea was logically implied by the principle of covariance of general relativity theory. For this principle automatically implies that if, from a particular reference frame, one type of force should discretely 'turn off', e.g. if the 'short range' nuclear force should cut off at a particular interparticle distance, and if, simultaneously, another type of force, e.g. the electromagnetic force, should 'turn on' there and then, then from the view of a different space–time frame of reference, these would not be the same physical conditions for turning one force law off and another on. That is to say, in one observer's frame, the nuclear force (in this example) would be in effect while from another frame of reference it would not be in effect.

The only way out of this dilemma would be to postulate that forces do not turn off while others turn on by the imposition of particular (non-covariant) physical conditions – such as the 'place' where the nuclear force of one particle turns on (or off) relative to another particle. What is implied here, as Faraday also concluded, is that forces of all possible types, solving all possible force laws, are always present, though they are not always seen to be manifested unless particular physical conditions are in place, according to the observer of these phenomena.

We see, then, that Einstein's quest for a unified field theory was not for purely aesthetic reasons or reasons of requiring simplicity. Rather, his investigation was to verify further the theory of relativity itself, since one of its logical requirements is the existence of a unified field theory of force, in the sense just described.

Finally, it might be asked: What logical ingredient is missing thus far from a truly unified theory in the Einstein sense, starting with the principle of covariance? As I see it, what is still missing is the explanation for the reaction of matter to the forces that are imposed on it. This reaction is called 'inertia'. Thus to complete the expression of a unified field theory of matter, it is necessary to set up a truly closed system, as implied by the full use of Einstein's field theory (as well as being implied by Newton's third law of motion in classical physics), so as to unify a generally covariant field theory of inertia with the unified field laws for the forces that matter exerts on matter – revealing both the action and reaction within the closed system, in terms of a covariant field theory[11].

In contrast with Newton's approach to inertial mass, whereby one inserts a mass parameter to quantify the amount of inertia that a body has, intrinsically, the covariant field theory of general relativity requires a field representation for the inertial manifestation of matter, derivable from the force fields that act on that matter. Indeed, this is the interpretation of the source of the inertia of matter, as due to its coupling with all other matter of a closed system, proposed by Ernst Mach and later named by Einstein: 'the Mach principle'. We will discuss this interpretation of inertia, that had a very seminal effect on Einstein's thinking, in Chapter 16.

With this introduction to the basis and motivation for Einstein's theory of relativity, we are now in a position to discuss the ideas of this very great contribution to our comprehension of the physical world and to our culture in a more explicit fashion. We will start in Chapter 2 with Einstein's initial impetus for his approach to this theory.

NOTES

1 A clear view of Galileo's derivation of his principle of inertia is given in: E. Mach, *The Science of Mechanics* (Open Court, 1960), p. 332.
2 Galileo's conclusion that all motion is subjective – leading to the idea that we now call 'Galileo's principle of relativity', is clearly explicated in: Galileo Galilei, *Dialogue Concerning the Two Chief World Systems*, Stillman Drake, transl. (California, 1967), p. 114, where Galileo says:

There is one motion which is most general and supreme overall, and it is that which the sun, moon, and all other planets and fixed stars – in a word, the whole universe, the earth alone excepted – appear to be moved as a unit from east to west in the space of 24 hours. Thus insofar as appearances are concerned, may just as logically belong to the earth alone as to the rest of the universe, since some appearances could prevail as much in one situation as the other.

3 For the views of Plato, Parmenides and Heraclitus, see: T. V. Smith (editor) *From Thales to Plato* (Chicago, 1956).

4 L. de Broglie, *Annales de la Fondation Louis de Broglie*, **4**, 56 (1979).

5 Discussions of the differences between the epistemological views of operationalism/positivism and realism may be found in standard texts on the philosophy of science. See, for example, E. Nagel, *The Structure of Science* (Harcourt, Brace and World, 1961).

6 G. Holton, in *Albert Einstein: Historical and Cultural Perspectives* (edited by G. Holton and Y. Elkana) (Princeton, 1982), p. xv.

7 In a letter Einstein wrote to David Bohm in 1953 (Einstein Archives, Jewish National and University Library, The Hebrew University of Jerusalem, Call No. 1576: 8 – 053), he said:

When one is not starting from the elementary concepts, if, for example, it is not correct that reality is described as a continuous field, then all my efforts are futile, even though the constructed laws are the greatest simplicity thinkable.

8 A clear discussion of riemannian geometry from the conceptual point of view, and its use in general relativity theory, is given in: C. Lanczos, *Albert Einstein and the Cosmic World Order* (Wiley, 1965).

9 The logical role of the 'principle of equivalence' in general relativity theory is discussed in: M. Sachs, On the logical status of equivalence principles in general relativity theory, *British Journal for the Philosophy of Science*, **27**, 225 (1976).

10 Faraday's views of a unified field theory are discussed in: J. Agassi, *Faraday as a Natural Philosopher* (Chicago, 1971); R. A. R. Tricker, *The Contributions of Faraday and Maxwell to Electrical Science* (Pergamon, 1966); M. Sachs, *The Field Concept in Contemporary Science* (Thomas, 1972).

11 This view has been pursued in detail in: M. Sachs, *General Relativity and Matter* (Reidel, 1982) and in: M. Sachs, *Quantum Mechanics from General Relativity* (Reidel, 1986).

2

A seminal idea – the principle of relativity

The initial impetus for the development of the theory of relativity in 20th century physics seems to have come when the curiosity of a 16-year old boy led him to ask a question about the nature of light. According to its most recent description, in the year 1895, light was known to propagate, maximally in a vacuum, at a speed of about thirty billion centimetres per second. Young Albert Einstein then wondered if it might be possible to see a light beam standing still, by travelling parallel to it at the same speed. To answer the question he had to determine the nature of the solution of Maxwell's equations for light, as described from a reference frame that moves parallel to a light beam at the speed of light. He anticipated that this analysis would then reveal the formal description of light at rest.

In contrast with his expectations – which he based on 'common sense' – Einstein discovered that, according to the form of Maxwell's equations (the most fundamental description of light known), if one insisted on maintaining the objectivity of these equations, then no frames of reference exist in which to describe light as propagating at any speed other than the magnitude of thirty billion centimetres per second! That is to say, even if he were travelling at this same speed parallel to a light beam, he should determine from his measurements that the light beam is moving away from him at thirty billion centimetres per second!

Of course, Einstein saw that he could avoid this seemingly nonsensical conclusion by assuming that the formal description of light in one frame of reference of space and time coordinates might be different from the way that light is described in different (relatively moving) frames of reference, as expressed from the first observer's view.

The following question then arose. Should the established 'common sense' notions be maintained, at the expense of altering a law of nature every time it is described (by a given observer) in a different frame of reference? Or rather, should the form of the law of nature be maintained in all possible frames of reference – even if this might be at the expense of violating 'common sense' notions? The initiation of the 20th century to the revolution of relativity was Einstein's acceptance of the latter alternative, asserting the objectivity of all laws of nature.

This choice then led Einstein to the generalization that all laws of nature, not only the particular law that governs the propagation of light, must have objective connotation i.e., the different expressions of a particular law of nature that underlie particular phenomena, as described in different space and time coordinate systems, must necessarily correspond with each other, in a one-to-one way, just as the words and the logical structure of the French sentence, 'le ciel est bleu', corresponds to those of the English sentence, 'the sky is blue', in accordance with the same physical content of the respective sentences in French and English. This is an assertion of the principle of covariance (also called the principle of relativity), that forms the axiomatic basis leading to all physical predictions in modern science attributed to the theory of relativity. The adaptation of this principle then led Einstein to introduce an alteration of the classical notions of space and time, and to the idea that the speed of light, which, according to this theory, is the maximum speed of propagation of any sort of force between matter, must be independent of the speeds of the interacting bits of matter, relative to each other.

It is important to emphasize that the theory of relativity does not imply a philosophy of 'relativism'. The latter is the belief that all truth is subjective, relative to an individual mind interacting with its environment. Such a view takes an anthropocentric view of the universe – a view that seems (to me) to have been rejected four centuries ago by the Copernican revolution. In contrast with the philosophy of relativism, the theory of relativity takes the view that there are purely objective (absolute) features of the real world that are independent of any individual mind that might be looking in!

One of these objective (absolute) features in the formal structure of this theory is the speed of light – the maximum speed of propagation of any force. Other objective features, as discussed in Chapter 1, are the laws of nature – expressions of order that imply cause–effect relations. Indeed, to ensure that any law of nature is objective, it became necessary in the theory of relativity to treat the space and time parameters as relative language elements, whose logical purpose is to facilitate a subjective expression for objective laws of nature. It is important to note here that the expression of a law of nature is not the law itself! The difference is the same as that between an entity's representation in terms of a word, such as 'love', and the essence of that entity – all of the emotions, oneness, beauty, anxiety, . . . that the word is supposed to represent. Finally, an objective feature of the universe is its own existence! For this is an entity that is, by definition, not relative to anything, since it stands for all that there is.

The classical view of space and time, rejected by Einstein's theory, was expounded by Isaac Newton, about 300 years earlier, and by the early atomists, such as the ancient Greek philosophers. Theirs was the idea that space and time have ontic significance – they are fundamental existents, independent of matter. With the classical view to describe a quantity of matter, and its action on other matter in terms of the motions so produced, one then relates the sequences of places and times occupied by some quantity of matter to its location within a fixed grid of points – the absolute space and time continuum. Einstein's alteration of this view was to use the space and time parameters as the subjective language elements to express objective laws of nature, and they were to have no more significance than this.

With this view, a very important purpose is served by the coordinate transformations in the theory of relativity. If the space and time parameters are only subjective language elements to be used in expressing physical laws, then to check if the expressions of an alleged law of nature in relatively moving frames of reference are indeed in one-to-one correspondence, as required by the principle of relativity, one must learn how to translate from the language of one of the reference frames to those of the others. The particular 'language transformations' that ensure the objectivity of the law of nature must then, apply to all other laws of nature. This is so because the transformations of coordinates only serve the role of translating one language into another. They are not, in themselves, physical cause–effect relations.

The first such coordinate transformations of the space and time coordinates, in relation to the theory of special relativity, were the *Lorentz transformations*, indicating how the space and time coordinates in Maxwell's equations, for electromagnetism, must change between frames of reference that are moving relative to each other at a constant speed in a straight line. This led to the theory of 'special relativity', as applied to all possible types of forces between matter. Einstein then recognized that with regard to the laws of nature and their general expression, there should be nothing unique about the comparison of laws in reference frames with this particular type of relative motion (called 'uniform motion'). He then generalized his theory to include the requirement of the objectivity of the laws of nature in frames of reference whose relative motion is not necessarily uniform – thereby leading to the theory of general relativity.

[It is interesting to note that this theory is usually referred to in the literature as 'the special theory of relativity' and 'the general theory of relativity'. It should rather, in my view, be called 'the theory of special relativity' and 'the theory of general relativity' – for it is the relativity that is special or general, and not the theory. Indeed, this is a single theory, the adjectives 'special' and 'general' referring to the relativity (of motion) and not to the theory *per se*.]

3

Early observations of things

According to the earliest writings concerned with man's attempts to understand the universe, there has been a captivation with the nature of things – their intrinsic constitutions and their motions. Why has there been such interest in the things of our environment? I think that the answer is probably rooted in our primitive instincts having to do with the drive for survival.

Of course, the survival instinct is shared by all living species – humans, fish, birds, insects, plants, . . . – but only the human being seems to have developed the very acute method to ensure survival having to do with the sense of reasoning. It has enabled humans to develop genetically to a more complex being – before we might have been eliminated by the elements – because it has enabled us to exert a mastery over other living species that outnumber us and have much more physical strength and endurance.

Along with the ability to reason, it seems that humans alone have developed the drive toward understanding the world, for the sake of understanding alone. That is to say, at some stage in our evolution, we reached a point of gaining personal satisfaction from the discoveries of abstract truths. This intellectual activity, which does not necessarily lead to material gains or other personal advantages, has, at each stage of our development, elevated the human race to such heights, and opened visions so fantastic and awe-inspiring, beyond the wildest dreams of our ancestors.

Since the days of antiquity, we were fascinated by the heavenly bodies in the night sky. Aside from the sheer beauty of the glorious distributions of shining objects that they saw, human observers started to wonder about the stars' shifting positions in the sky, from night to night – why some of them hardly moved at all in one night and why some of them seemed to move across the entire sky. What were those very swiftly moving objects that seemed to have a tail – like a cosmic snake? What was that large silvery disc that seemed to change its shape from a cradle to a full circle, throughout the month? What was all of this activity that was going on in the night sky?

The Greek philosopher, Plato, in the 4th century BC, speculated that the paths of the heavenly bodies must necessarily be circular. For a circle is the most beautiful of all forms – it looks the same from any angle, as viewed from its centre; it has no beginning and no end. Plato also believed that the universe is eternal – that it has always existed and that it will continue to exist forever, though it has had different forms in the past, and it will continue to shift its form in future stages of its existence. He applied this view to all substates of the universe, including human beings, thereby leading him to a belief in our immortality – through the process of reincarnation. Plato also believed in the fundamental oneness of the universe – a totality without actually separable parts. The four elements to Plato – air, earth, fire and water – were only mathematical forms, representative of the ideas in the minds of human beings that give an image of the real structure of the world[1].

Plato's student, Aristotle, agreed with him that the world is constituted of these four elements; though he saw them as four real parts, rather than mathematical forms. Aristotle's view of the world in terms of its concepts, was more materialistic than abstract[2].

Aristotle believed that our planet Earth must be at the centre of the universe because, he speculated, of the four basic elements, Earth has such a nature that it must be down, while the heavenly bodies, being made of the other elements, are up. He then concluded that Earth must be at the centre of a spherical universe. A few hundred years later, the Greek astronomer, Ptolemy, mapped the observed paths of the stars and planets in such a way as to verify this geocentric model of the universe.

The geocentric cosmology of Aristotle was in constrast with the earlier cosmology of Pythagoras, several centuries before him, in Greece, who speculated about the cause of the motions of the objects of the sky – that there must be a 'central fire' at the origin of the spherical universe, causing all of the objects of the sky, including Earth, to move[3]!

This dispute about whether or not the Earth 'moves' did not reappear again in the history of ideas until the Copernican revolution, 17 centuries later, particularly the trial of Galileo by the Inquisition of the Christian Church – charging him with heresy for his claiming that the Earth moves[4]!

In addition to Aristotle's conception of matter, he also had interesting views of space and time. He defined space as that which is occupied by matter. That is to say, space, to Aristotle, is non-existent without matter to fill it. Because he claimed that the amount of matter in the universe must be finite, he concluded that, correspondingly, the space of the universe must be finite. That is, since his observations of the night sky seemed to indicate that the matter of the sky is bounded in a fixed proportion, then space itself must be limited.

Aristotle defined time as a measure of motion. That is, without the motion of matter, there would be no time to talk about! He further speculated that all observable motions must be derivable from other motions that preceded them. That is, if a particular object at rest should start to move at some instant in time, this must be caused by an 'accidental' transfer of motion from some other matter. (Perhaps this idea corresponds to the law of 'conservation of momentum', discovered centuries later in the history of physics?) Aristotle then argued that this succession of the causes of motions logically implies that potential motions extend to the infinite past and will extend to the infinite future. He then concluded that time, as a measure of motion, must be unbounded, i.e. it is infinite in extent.

In anticipation of the later interpretations of time, that will be discussed in further chapters, it is important to note that Aristotle's meaning of 'motion' is an abstract, continuous entity. To him, it is the 'potentiality' for matter to move. The actual motion of matter that we experience, in Aristotle's terminology, is called 'locomotion' – the 'actuality'.

Though 'motion' and 'time' are abstract entities to Aristotle, they would be undefined concepts, from his view, in a matterless universe. This is because, logically, potentiality is a meaningless concept without the existence of a material entity to actualize it. Later on in the history of ideas, Faraday's 'field of force', corresponded in a way with Aristotle's potentiality, and Faraday's role of the 'test body', an entity to be influenced by the force field, is similar to Aristotle's actuality of motion – his 'locomotion'. This important development in 19th century physics, and its role in the discovery of relativity theory, will be discussed later[5].

A question that naturally arose in the days of antiquity concerned the causes of the motions of the objects of the night sky. Why did they move at all? What caused some of them to move more swiftly than others? Aristotle believed that things move at one speed or another because of where they are in space and when they are there. He speculated, then, that space has causal powers and that things move at one speed or another because of an intrinsic impetus that forces such motion. The existing motions are not due to the presence of other matter, but rather, they are due to their locations in space. For example, an apple falls from a branch on a tree to the ground because its natural place in space is 'down' rather than 'up'. But once the external (spatial) cause is no longer present, in Aristotle's view, the matter would remain at rest, as its natural state.

It wasn't until about 1700 years later that Galileo, in the 16th century, learned that a body will move naturally at a constant speed, forever, without any need for an impetus, or what was later called a 'force', to maintain this state of motion. Instead, it was discovered by Galileo, and a generation later by Newton, that the effect of a force was to change the state of constant speed (or rest). This idea[6] is called Galileo's principle of inertia. A contemporary of Galileo, Rene Descartes, also deduced that constant motion is a natural quality of free matter.

Aside from these motions of the things in the sky, precisely what is that nature of their constitutions? Are they, perhaps, the eyes of demons looking down on the human race, controlling its destiny? And what about the common, everyday matter that we see here on Earth, such as a stone on the side of a hill, or the fishes that fill the oceans and lakes? Is this matter really composed of combinations of the four basic elements: air, earth, fire and water? From these times in ancient Greece to the period of Galileo and Newton, most people believed that the material world must be made up of distinguishable things, each moving along its own trajectory in space. The way to describe these things is in terms of their external physical properties, such as shape, size, colour, etc. and in terms of their respective locations at different times. It has also been believed since the earliest times that the ordinary matter, that we respond to, is made up of more elementary units – the 'atoms' which supposedly arrange themselves in different sorts of patterns to give the outward characteristics of the observed qualities of matter. Such a view was promoted in ancient Greece by several scholars who preceded Plato and Aristotle, most notably Democritus. Not everyone accepted this atomistic view of the material world. The alternate view was that matter could be divided up continuously, without limit. But the majority of people, to this day, have accepted the atomistic approach, in one form or another.

In the 17th century, Isaac Newton was a strong adherent of the view of atomism. His law of universal gravitation implied that, generally, forces are the exertions of an action of one quantity of matter on another, whose strength depends only on their mutual separation. That is, the action of one body, A, on another, B, somewhere else, was taken to be a function of the existence of A at its location in space, relative to the location of B. Should A then be extinguished, or moved to a different location, B would react to this change spontaneously; that is, without any time delay. This concept is commonly called 'action-at-a-distance'.

Newton strongly believed that the same fundamental principles that underlie an explanation of gravitational phenomena must underlie all other physical phenomena as well, such as optics, electricity and magnetism. Thus he devised a particle theory of light, and with this theory he was able to explain most of the known optical properties (of his day) in terms of the attraction and repulsion between the particles of light and the particles of ordinary matter[7]. Still, one optical phenomenon that Newton's theory of light could not quite explain was the 'diffraction effect' – making it appear that light can 'bend around corners' and interfere constructively and destructively, as waves combine.

Some of the electric and magnetic manifestations of matter had been discovered by the time that Galileo and Newton appeared in the history of physics, but the great flurry of experimental and theoretical study on the nature of these phenomena did not come until the scientific investigations in the 18th and 19th centuries. One of the outstanding investigators of this period was Michael Faraday. His achievement was not only to show that the force of electricity and the force of magnetism are, objectively, a unified electromagnetic force. More generally, he also introduced a conceptual revolution in the way of understanding matter in fundamental terms. At the expense of abandoning Newton's atomistic concept, based on the notion of 'action-at-a-distance', Faraday proposed that underlying outward observations of things, there was a more primitive building block in the nature of matter, described as a continuous field of potential force. This was taken to be an entity not directly observable, yet at the root of all material properties. He then took a fundamental stand in terms of abstract rather than directly perceivable entities, as the fundamental starting point for a theory of matter. Indeed, this approach served as a very important precursor for the development of the theory of relativity in the 20th century.

NOTES

1 Plato expressed his philosophy of the cosmos primarily in his dialogue, *Timeus*. See: *Plato, Timeus and Critias*, D. Lee, transl. (Penguin, 1965).
2 A classic translation and interpretation of Aristotle's works is: R. McKeon, *The Basic Works of Aristotle* (Random House, 1941).
3 W. K. C. Guthrie, Pythagoras and Pythagoreans, ın his *History of Greek Philosophy* (Cambridge, 1962).
4 There is a wealth of literature on the conflict between Galileo and the Christian Church. See, for example, G. de Santillana, *The Crime of Galileo* (Chicago, 1955); P. Redondi, *Galileo: Heretic* (Princeton, 1987).
5 M. Hesse, *Forces and Fields* (Nelson, 1961); M. Sachs, *The Field Concept in Contemporary Science* (Thomas, 1972).
6 A. R. Hall, *From Galileo to Newton* (Dover, 1981).
7 I. Newton, *Optiks* (Dover, 1979).

4

Toward an abstract view of nature

In the 18th century, a Jesuit priest, Roger Boscovich, introduced a new approach to natural philosophy. He took the continuous manifestation of matter, that is its power to act on other matter in space, to form a fundamental starting point for a description of the material universe[1]. About 100 years later, Michael Faraday adopted this view of the continuous field concept to describe electricity and magnetism most primitively.

According to Faraday, rather than looking upon the potential field of force that could be exerted by a bit of matter on other matter (should the latter be located at any of the continuum of spatial points) as a secondary, derivative property of that matter, one should rather consider the continuous field of potential force as fundamental. He then viewed the 'discrete particle' aspect of matter as derivative. According to the field theory proposed, the real stuff of the material world is the abstract (i.e. not directly observable) aspect associated with the potential field of force of matter. This view challenged a prevailing philosophic stand, presently known as 'naive realism', which asserts that: only that which we human beings directly perceive to be there, outside of us, is the reality from which a true description must follow[2]. Faraday's abstract approach, on the other hand, took the fundamental reality to be at a level underlying that of human percepts. He might have asked: After all, why should the human mind, in particular, be so designed that it can respond directly to the truths of nature? Perhaps to discover nature's secrets, we should rely most heavily on our ability to reason, using the information given to us by means of our percepts as not more than hints about the real world.

The electric field was conceived by Faraday as a continuum of 'lines of force', each line starting at a positive charge and ending at a negative charge. A positively charged 'test particle' may then be defined as that sort of object that would move from the positive end toward the negative end of the electric line of force. However, to be logically consistent, the 'test particle' also must not be thought of, in fundamental terms, as a discrete particle. The feature of spatial localization is rather a secondary aspect of even the test body – a feature which is exploited when investigating the nature of the continuous field of potential force of other matter.

An interesting implication of this explanation of electricity is that for every positively-charged matter component of a system, there must exist a negatively-charged matter component. This is simply because a line has two ends! With this view, it would then be as meaningless to introduce a negative charge without the existence of a positive charge as it would be to talk about 'up' without the existence of 'down'.

The magnetic lines of force have a different shape from their electric counterpart. Rather than terminating on oppositely polarized 'ends', the magnetic field lines close on themselves. To describe the magnetic force in terms of 'north' and 'south' poles, one must consider them together as a dipole, with infinitesimally small extent in space. Again, according to Faraday, it is the continuum of such lines of force, this time closing on themselves, rather than the magnetic dipole (the 'thing'), that is fundamental in the description of magnetic forces. In this case, the 'test body' that is to probe the magnetic field of potential force, is a magnetic dipole itself, such as a compass needle. The orientation that this dipole is forced into, along the curved magnetic field that is probed, then determines the orientation of this field – the north pole of the compass is attracted to the south pole of the probed field and vice versa.

The very important feature of electrically charged matter that Faraday discovered, which led directly to the theory of relativity, was that a magnetic field of force is nothing more than an electric field of force in motion, and reciprocally, an electric field of force is nothing more than a magnetic field in motion. Since one can always find some frame of reference that is in motion relative to any other one, such as the frame of the scientist's chair, that may for convenience be considered as fixed, relative to something that may be moving in an experiment, relative to her chair, it then becomes artificial to talk of a 'pure' electric force or a 'pure' magnetic force, objectively. For, the experience of an electric force to one observer of, say, a static, electrically-charged pith ball, would be experienced as a combination of electric and magnetic forces to a different observer, who would be in motion relative to the pith ball. Thus, Faraday discovered that the unification of electricity and magnetism into electromagnetism is in fact a consequence of the full description that incorporates motion, and therefore relates fundamentally to the incorporation of space and time. This is a description of electromagnetism that has the same general form in all possible relatively moving frames of reference, from the view of any particular observer who compares the expressions for the law of electromagnetism in these frames. The latter is a statement of the principle of relativity, that we have discussed in Chapter 1, as the fundamental basis of the theory of relativity[3].

During the time that Faraday was doing his researches on electricity and magnetism (at the Royal Institution, in London), a younger theoretician,

James Clerk Maxwell, accepted a professorial chair at King's College in London. The subsequent meeting of these scientists in London led to some glorious consequences for physics. Maxwell learned how to express Faraday's results in a more precise mathematical form – 'Maxwell's equations' – whose solutions then revealed all the features of a unified theory of electromagnetism, as well as all the features of optics and predictions of new phenomena, such as radio waves. The latter were discovered in experimentation not long afterwards, by Heinrich Hertz, in Germany[4].

Maxwell's equations, as the form of the law of electromagnetism, was more complicated than the equations for gravitation, given by Newton's law. The extra complication came in because, unlike the gravitational force which depends on only one independent variable – the distance between interacting bodies – the electromagnetic force depends on four independent variables – the three spatial coordinates, separately, and the time parameter. The complication comes in because an electrical charge in motion, say in the x-direction, relative to another direction that locates a test body, say the y-direction, results in a reaction that the test body is caused to move in a third direction – along the z-axis. Or, the direction in which a force acts on a test body, as caused by the motion of this body through an external magnetic field (say near a bar magnet), is oriented perpendicular to the plane formed by the directions of the external magnetic field and the direction of motion of the test body.

We see, then, that Maxwell's equations depend on the variations of force fields in the four dimensions of space–time. These are partial differential equations, while Newton's equation of motion for universal gravitation is an ordinary differential equation, depending rather on a single independent variable – the distance between the interacting masses.

After discovering how to formulate the equations whose solutions would correctly describe all known manifestations of electricity and magnetism, Maxwell discovered that there are additional solutions, corresponding to the very-far-away effect of charged matter, which imply that a test charge would oscillate in a plane perpendicular to its direction from the source of these electromagnetic fields, at any one of a spectrum of frequencies. Such solutions of Maxwell's equations, called the 'radiation solutions', precisely describe the effects associated with (monochromatic) light, i.e. light of any one of a spectrum of frequencies, as well as other frequency solutions, later found to be identified with radio waves, X-rays, gamma rays, etc.

According to the structure of these solutions, the speed of propagation of the radiation waves, for all possible frequencies, is precisely the same as the speed of light in a vacuum.

At first, this speed appears in the formalism as the ratio of magnitudes of the electrical to the magnetic fields of force that would exert the same quantity of force on a test body. This speed appears in these equations in the first place because, basically, these laws unify electricity and magnetism, wherein the units of these fields, called $\mathbf{E}$ and $\mathbf{B}$ are different. It is necessary to express them in the same units in order to unify them into an electromagnetic field, so a universal constant is introduced with the dimension of speed – which turns out to be the speed of light when the solutions of the theory are compared with the empirical data. That is, when this speed is numerically equal to the speed of light, c = thirty billion centimetres per second, then the field intensities, $\mathbf{E}$ and $c\mathbf{B}$ would accelerate any test charge at the same rate.

It then followed from combining this result with the known properties of light – e.g. its feature of being a transverse vibration, the relation of its frequency spectrum to the colour spectrum of the rainbow, etc. – that all known optical phenomena were nothing more than the propagation of mutual electromagnetic forces between distant quantities of electrical charged matter, at the speed c.

The ground was then ripe for the appearance of the theory of relativity. For the hint was already in the structure of the electromagnetic equations that in our search for scientific truth, perhaps we should demand that any law of nature should have the same form to any observer who may deduce this law, irrespective of his or her motion relative to any other observer, and irrespective of the particular language that may be used to express this law – be it English, Martian, beetle language, or the relative measures of space and time to observers in relatively moving coordinate frames.

Question What is the connection between what you call an 'abstract approach' to the laws of nature, on the one hand, and the way in which you introduce the space and time coordinates into the natural laws, on the other?

Reply With the abstract approach, we start with some general assertions to underlie a set of ordered relations that are supposed to describe correctly some manifestation of nature. These ordered relations are expressed by us in terms of a particular language composed of words and a logic that connects them. We have found that the most precise sort of language to use is that of a continuum of space–time points and their logic. It is important to note that this continuous set of parameters are the elements of a language, in this view, i.e. they are not in themselves physically real, objective entities. It follows from this idea that if other members of a physical system, say intelligent beetle-like creatures from Galaxy-X, should be in a position to communicate laws of nature, their expression of these laws would not be in terms of the space and time 'words'. But if they are indeed expressing the same truths as we are, about similar phenomena, then there should be a one-to-one correspondence between their expression of these truths and ours – according to Einstein's principle of relativity (principle of covariance).

Of course, if we could not communicate with these beetles, we would never know whether or not our respective expressions for the same laws of nature correspond. To check if they are so, we would have to find the translation between our respective languages.

With this approach, we would have to say that there is nothing special about the world because of the way in which human beings, in particular, see it! The human being uses his perceptions to arrive at abstract laws of nature, as far as he can. These are, by definition a set of logically ordered relations, giving him some of the objective features of the universe. That is, the relations that we seek are those that are independent of whether or not we are there to perceive the manifestations of the universe.

Question Then is this a fundamental change in the way that we see ourselves in the universe?

Reply With the earlier (atomistic) view, our thoughts about the world, especially in the Western philosophies, were largely guided by an empirical approach[5]. It implied that the way the world is depends on how we see it! But then, with the coming of the Copernican revolution, this view was gradually replaced with less and less of an anthropocentric model[6], until the arrival of the (partially) abstract field concept of Faraday, and the (totally) abstract field concept of Einstein.

With the abstract view of relativity theory, humankind is not taken to be at the centre of all that is knowable. Indeed, we are not more than an infinitesimal manifestation of the universe, occupying a position in it that is not any more special than any other position. As human beings, we must acknowledge the gift of a faculty to think rationally, in addition to our perceptive capabilities. We then have this facility to probe the universe by using our sensations, as hints, to help in a rational analysis of the real world. This is indeed our obligation as scientists and philosophers. This approach to knowledge is called realism[7].

NOTES

1 R. Boscovich, *A Theory of Natural Philosophy* (MIT, 1966).
2 Einstein quotes Bertrand Russell's refutation of 'naive realism' in: P. A. Schilpp, (editor), *The Philosophy of Bertrand Russell*, (Harper and Row, 1944), p. 281, as follows:

> We all start from 'naive realism', i.e. the doctrine that things are what they seem. We think that grass is green, that stones are hard and that snow is cold. But physics assures us that the greenness of the grass, the hardness of stones, and the coldness of snow are not the greenness, hardness and coldness that we have in our own experience, but something very different. The observer, when he seems to himself to be observing a stone, is really, if physics is to be believed, observing the effects of the stone upon himself. Thus science seems to be at war with itself: when it most means to be objective, it finds itself plunged into subjectivity against its will. Naive realism leads to physics, and physics, if true, shows that naive realism is false. Therefore, naive realism, if true, is false; therefore it is false.

3 Faraday's researches on electricity and magnetism are reported in his original writings in: M. Faraday, *Experimental Researches in Chemistry and Physics*, I, II, III (Taylor and Francis, 1859).

4 A recent publication that reports all of Mawell's researches is: J. C. Maxwell, *A Treatise on Electricity and Magnetism*, (Dover, 1954), 3rd edn, Vol. 1, 2.

5 A clear discussion of the concept of atomism is given in: B. Russell, *On the Philosophy of Science* (Bobs-Merrill, 1965), Chapter 1.

6 In recent times, a literature has developed on a return to anthropomorphism – the view of a human-centred universe, in various versions. One version, called the 'weak anthropic principle', asserts that the existence of human beings is consistent with the physical properties of the universe as a whole. A second version, called the 'strong anthropic principle', asserts further that the universe must have physical properties that allow human beings to come into existence, at some stage of its development. These views of the anthropic principle are expounded in: J. D. Barrow and F. J. Tipler, *The Anthropic Cosmological Principle* (Oxford, 1986). They are further discussed in the recent book by S. W. Hawking, *A Brief History of Time* (Bantam Books, 1988), p. 124. It is my view that the anthropic principles, whether in the 'weak' or the 'strong' version, are a regression to the pre-Copernican views of the universe, and the human being's place in it – views that were in place more than 500 years ago! – before the progress that was made by Copernicus and his scientific and philosophical successors.

7 A discussion of the philosophic view of 'realism' is given in: R. J. Hirst, Realism, in *Encyclopedia of Philosophy* (edited by P. Edwards) (Macmillan, 1967), Volume 7, p. 77. My own understanding of the view of 'realism' is the same, I believe, as that of B. Russell and A. Einstein – the idea that there is a real world, irrespective of whether or not there may be human observers of it, and what it is that they may perceive.

5

Einstein's ideas of special relativity

As we discussed previously, Einstein discovered the theory of special relativity when he tried to answer the following question. Would it be possible to find a frame of reference of space and time coordinates, moving along with the speed of propagating light, such that the light would be seen to be stationary from this frame? From a close examination of Maxwell's equations – the physical laws that explain light – Einstein discovered that indeed these equations have no solutions, expressed in any frame of reference, in which light would be at rest, or propagating at any speed other than the speed c (thirty billion centimetres per second) – the speed of light in a vacuum. (Recall that light propagates at slower speeds than this in material media. This is because of the delaying action of its being absorbed in the matter, thereby increasing the energy of this matter as it becomes 'excited'. The light is then emitted again later, when the matter returns to its ground state – i.e. when it 'de-excites' in the characteristic 'lifetimes' of the light-conducting matter.)

Einstein's conclusion about the constancy of c in all reference frames was based on his tacit assumption that the form of Maxwell's equations must be preserved in all relatively moving reference frames. He was not willing to give up this assumption – even to preserve common sense – because it seemed basic to the meaning of the phrase 'law of nature'. Thus he was led to a shocking conclusion that indeed defied common sense.

To exemplify how 'far out' his conclusion was, especially for the thinking in science in the beginning years of the 20th century, compare it with a typical 'common sense' situation. A girl standing by a railway track watches a train pass by at 60 mph. She observes a man standing on the moving train, throwing a ball in the direction of the train's motion, at 10 mph, relative to the roof of the train. One would then predict that the girl should see the ball moving at a speed of (60 + 10 =) 70 mph relative to the ground, in the direction of the train's motion. Common sense tells us that this must be so! On the other hand, if the man on the train had ejected a

light beam instead of a slowly moving ball, then since the light travels at
near 669 600 000 mph, the girl should expect to see that, relative to the
ground, the light would be moving at (669 600 000 + 60 =) 669 600 060 mph.
Yet, according to Einstein's theory, the girl would see the light propagating
at 669 600 000 mph relative to the ground – as though the emission of the
light from the moving train, rather than the still ground, was of no
consequence! That is to say, the measurement made by any observer should
reveal that the speed of propagation of light is totally independent of the
speed of the source of that light!

This conclusion follows if we are to insist that the law of nature that
underlies an explanation for the phenomenon of light must have the same
form in any relatively moving frame of reference, from any particular
observer's view. While this assertion may be very easy to believe, indeed
it seems to be self-evident (in the definition of the word 'law'), the further
conclusion that the speed of light is independent of the speed of its source
seems to violate common sense. Still, its verification is well established in
the successes that Maxwell's equations have exhibited in correctly predicting
features of light (as well as all other electromagnetic phenomena) –
maintaining their form in all relatively moving frames of reference.

Early statements about the axiomatic starting point of the theory of
relativity asserted that it was based on two postulates:

1 the principle of relativity, asserting that the forms of all of the laws of
   nature are independent of the frame of reference of space and time
   coordinates in which these laws are expressed; and
2 the maximum speed of propagation of interactions (the electromagnetic
   interaction in the case of light) is a universal constant – i.e. it must
   have the same value as determined from any frame of reference, in all
   other relatively moving frames of reference. This constant has been
   established to be numerically equal to the speed of light in a vacuum.

In the early version of the theory, the different frames of reference were
distinguished from each other by virtue of their relative motion of a special
type – that which corresponds to a constant relative speed in a straight line.
Such coordinate systems are called 'inertial frames of reference', and this
version of relativity theory, where the forms of the laws of nature are
preserved with respect to transformations between inertial frames of
reference, is called 'the theory of special relativity'[1].

These two postulates, asserted by Einstein in the early stages of his
theory, then generalized his conclusions about electromagnetic theory in
particular. That is, Maxwell's equations are a particular law of nature having
to do with electromagnetic phenomena. But Einstein's first assertion is that
all laws of nature must each have the same form in any relatively moving
frame of reference – this is the 'principle of relativity' (or, 'the principle of
covariance'), that we discussed previously, in the Introduction.

It should be noted at this point that implicit in the principle of relativity is the tacit assumption that there must exist laws of nature in the first place – logically-ordered relations that are supposed to underlie explanations of physical phenomena. As it is with the human being's limited reactions to his or her environment, the minimum number of variable parameters ('words') that are needed to express these laws is four. These are usually identified with the three spatial measures and one time measure – that we correlate with measures of extension and duration. Thus, to check if the principle of relativity is indeed valid in regard to an alleged law of nature, it is necessary to compare the expressions of this law in the relatively moving frames of reference, from the view of any particular observer.

To express the laws of nature, from a particular reference frame, one may use the 'words', x_1, x_2, x_3, to locate a point in 3-dimensional space, and t to refer to a particular measure of time. In another frame of reference, that may be in motion relative to the first one, to express a law of nature, the observer from the first frame would use the 'words' x_1', x_2', x_3', t'. To compare the respective expressions of the given (alleged) law in these frames, it is necessary to find the correct 'translation' of the words of the space–time coordinate systems,

$$(x_1, x_2, x_3, t) \rightarrow (x_1', x_2', x_3', t')$$

The translation of the space and time 'words' used in one reference frame into those of other reference frames that move relative to each other at constant relative speeds in straight lines, that would force the respective expressions of the physical laws to be in one-to-one correspondence in all of these (inertial) frames, were discovered in regard to the Maxwell equations by H. A. Lorentz – these are the 'Lorentz transformations' – underlying the covariance of the laws of nature according to special relativity.

Lorentz did not interpret his transformations in the same way as Einstein. His interpretation depended in particular on the nature of light and the way it propagates through (his assumed) aether medium – the assumption about the need for an aether medium to conduct light was made by most of Lorentz' contemporaries and predecessors, such as Maxwell and Newton. Einstein, on the other hand, did not attribute physical connotation to these transformations other than changes in the 'language' of space and time coordinates that would ensure the objectivity of the laws of nature in all reference frames. Thus he had no need for the model in which light must propagate through an aether medium; that is, to Einstein, the aether was a superfluous concept.

Special relativity, then, refers to the objectivity (covariance) of all laws of nature with respect to their comparison in all inertial frames of reference – frames that move with respect to each other at constant relative speeds in straight lines. For example, two aircraft that fly at the same elevation at constant relative speeds, at different locations near the Earth's surface, are inertial, relative to each other, but they are not inertial relative to the

Earth – because of its gravitational force that acts on the aircraft – that would make them accelerate downwards toward the Earth's surface, if not for the counteracting force that keeps them in a fixed orbit. The significance of inertial reference frames is that one cannot distinguish forces relative to each of them, since the latter is a cause for accelerated motion (Newton's second law). In any case, the theory that asserts the objectivity of all laws of nature in regard to their comparison in inertial reference frames is 'the theory of special relativity' – first enunciated by Einstein in 1905[1].

There has been a question about the separation of the principle of relativity and the universality of the maximum speed of propagation of interaction (the speed of light in a vacuum), as logically independent axioms of the theory, or whether the second of these axioms is logically derivable from the first. We saw in the Introduction that, indeed, the universality of c is a logical consequence of the principle of relativity (covariance), because of the unification of space and time into space–time in this theory.

Question   I find it difficult to understand, logically, how the speed of light, or the speed of anything else, can actually be independent of the speed of the source that emitted it, relative to a stationary observer. Could you say anything more about this?

Reply   I understand your difficulty. It has to do with a violation of your common sense notions – your 'intuition' about how motions should add up. But common sense, after all, is not more that a feeling that you develop, based on your accumulated experience. I don't think that you ever had the experience in which your velocity relative to Earth may have approached the speed of light. Based on a purely rational analysis, the only test of our assertion about the speed of light propagation is whether or not the theory that is formed, based on this idea, can make a series of other predictions not made by other theories, that agree with the empirical facts. If this theory would do this, then one would have to agree that so far as we know, the theory that incorporates the idea of the universality of the speed of light is scientifically valid.

I think that a second difficulty in your thinking has to do with your (perhaps subconscious) insistence that light is a 'thing', like a piece of rock or a jet plane, hurling through space on its own. I will discuss, later on, the idea that this may be a faulty interpretation of light; that one should rather view light as a mutual relation between electrically coupled matter – an interaction that propagates between them at a speed that is independent of the speeds of the coupled, electrically-charged matter, relative to each other, but is indeed a feature of their law of coupling, and not a 'thing-in-itself'[2].

NOTES

1 The original version of Einstein's theory of special relativity is his 1905 article: On the electrodynamics of moving bodies, in: A. Einstein, H. A. Lorentz, H. Weyl and H. Minkowski, *The Principle of Relativity*, W. Perrett and G. B. Jeffery, transl. (Dover, 1923).

2 This view is consistent with the 'delayed-action-at-a-distance' concept. In the context of a theory of 'all particles and no fields', it was developed in: J. A. Wheeler and R. P. Feynman, Classical electrodynamics in terms of direct inter-particle action, *Reviews of Modern Physics* **21**, 425 (1949). In the context of 'all fields and no particles', it was developed in this author's theory in: M. Sachs, *Quantum Mechanics from General Relativity* (Reidel, 1986), Section 7.5; it was discussed further in: M. Sachs, *Einstein Versus Bohr* (Open Court, 1988), p. 229.

6

Space

Should the average person be asked: What, precisely, do you mean by the word 'space'?, the answer might be: Space is, for example, the inside of an empty drawer – a region that is there to be filled with socks and shirts. Generally, then, space is seen as an entity to be occupied by matter; i.e. space and matter are quite independent of each other. The continuum of points of space are like a continuous sequence of rungs of a ladder. Any one of them might be occupied by a person's foot, but it need not be so. The ladder is still there, with or without a person to climb on it.

Recall that Aristotle's view was that space cannot be defined without matter to occupy it. To him, there was no sense to the notion of a 'vacuum'. But one of the most significant philosopher–scientists, Sir Isaac Newton, did take the view that there must indeed exist an absolute space, in which things are located and move about from place to place. Of course, since our percepts respond only to the effects of matter on us, (and not to 'empty space') we must deduce that the concept of a vacuum i.e. 'empty space', is something not directly observable. However, so long as this entity would have observable consequences, it should still be an important ingredient in the description of the material universe. Those who rejected the concept of absolute space, such as Newton's contemporary, G. W. Leibniz, did so only out of the conviction that there could be no observable consequences of an absolute space. [This question is still being debated today, in physics by the cosmologists[1].]

Of course, as human beings, we do have a physiological reaction in the optical sections of our brains, that we interpret in terms of a three-dimensional space about us. We call this 'visual depth perception'. We also have a physiological reaction in the auditory section of our brains that gives us the 'stereophonic' sensation of three-dimensional space. In addition, our built-in equipment responds optically and acoustically to 'distance' from a source – the more intense the light or sound, the closer we judge its source to be from us.

Still, it must be admitted that all of these conclusions about the three-dimensionality of space are indeed our mind's interpretations of sensations that occur at only one spatial location – within the perceiver's 'brain'. [Though the nerve centre of the body is indeed in the brain, is it possible that the sensations in our thinking, i.e. in our minds, actually entail our entire bodies, holistically, rather than only our brains?[2]]

Is it possible, then, that this is a faulty interpretation about three-dimensional space, out there? Is it truly valid to draw the conclusion about a one-to-one correspondence between the human being's brain reactions and a three-dimensional space that is independent and wholly outside of the human respondent?

The philosophic stand of 'idealism' and that of 'naive realism' lead to opposite answers to this question. A well-known proponent of the approach of idealism was George Berkeley[3], who argued that indeed one has no justification for concluding the existence of any reality outside of his individual self. He claimed that only God, and the entities not subject to the laws of physics, such as the souls of human beings, can be asserted to exist outside of the human organism. This view totally supported the idea of the separation of mind and matter – a conclusion that has been debated pro and con for centuries.

A leading supporter of such a separation was the great rationalist, Rene Descartes[4]. A leading contender for the philosophy that there is no such fundamental separation was Baruch Spinoza[5]. Descartes' philosophy advised the natural philosopher to begin in the search for truth by dismissing all prejudice – by doubting all ideas in our (preconceived) knowledge. He then asked himself: What do we really know? He answered that we certainly know that we exist, because of our thinking this question – *cogito ergo sum*! His next question (an entrance into the new field of psychology) was: What is it to think? Is this a mechanical process in which substantive forces are exerted by matter on matter? Descartes answered that the action of the mind on the body is certainly not a mechanistic process between coupled material components. He saw mind rather as an entirely different sort of conceptual entity from matter.

Since mind has none of the materialistic attributes of matter, according to Descartes' speculation, mind and matter cannot interact with each other, at least in any direct way. With this conclusion, Descartes was then faced with the dilemma of explaining the apparent fact that our thinking has an influence on our physical behaviour. For example, one 'sees' a piece of chocolate (body), one then 'thinks' about how delicious it would be (mind) – causing one to then reach for it (body).

Descartes concluded that while the actions of the body and the actions of the mind are not in any way physically dependent on each other, in a material, causal fashion, there must be a one-to-one correspondence between each elementary mental process and each elementary physical process of the body. Yet, he felt that it is necessary, according to his speculations on this problem, to postulate that both processes are like trains moving along parallel tracks, not influencing each other's motion in terms of cause–effect relations, yet describing parallel behaviours that are in correspondence with each other.

Descartes then agreed with the notion that mind and matter are separate, independent entities. But he did not agree with Berkeley's contention that the reactions of our senses (thinking) have no relation to reality. For Descartes assumed that there indeed exists a real correspondence between the processes of the mind and processes of matter. That is, if a real flower should be brought close to one's nose, the response in the thought process is a reaction that corresponds to that real flower.

Spinoza did not agree with Descartes' conceptual separation between mind and matter. Instead, he proposed a philosophy that implies that any apparent separation between mind and matter must be illusory. Spinoza started with the assumption that the real world, including the human race, is fundamentally a closed system – manifesting itself in one way or another, depending on corresponding sets of correlated conditions[6].

In the Spinozist view, one of the manifestations of this single, closed universe, is the human race itself. According to the Spinozist view, then, the universe is a single, substantive entity, characterized by a set of logically connected relations – representing its underlying order. No manifestation or aspect of the universe is independent of the rest of the closed system. It then followed from these speculations that the characteristics of mind and those of matter are indeed not independent of each other. They are, rather, distinguishable manifestations of a single, closed system, that is in fact without parts, just as the different notes sounded by a violin are not more than the manifestations of the whole violin. This is the philosophic approach of 'holism'.

With this view, the logical relations that underlie the order of the universe are like the forces that produce the tensions in the violin string, the coupling of the vibrating string to the surrounding air molecules in causing them to vibrate at particular frequencies, the coupling of the longitudinally oscillating air molecules to an individual person's ear drums, eventually resulting in the sense impressions in the brain that are identified by the mind with the notes sounded by the violin.

It will be seen, later on, that this view of the universe, as a closed system, is also implied by a full exploitation of Einstein's theory of relativity. Einstein's view, in turn, when fully generalized in the form of general relativity theory, comes close to the Spinozist concept of the universe.

The view of naive realism in natural philosophy contends that all that a human perceives with the five senses, outside of the body, is truly there in reality, in its raw, fundamental form. With this approach, what we humans react to in the material world is *it* – a set of physical characteristics not reducible to any more primitive (metaphysical) explanation.

According to the view of idealism, our conclusion that there is indeed a three-dimensional space that is real, because of our perceptions of it, is no more than an illusion. With the view of naive realism, on the other hand, the three-dimensional space is there in reality, just as we perceive it, but independent of our being there to perceive it! This philosophic approach contends that this is so, simply because we are there to perceive it to be so[7].

For a time, many felt that these were the only two possible philosophic views of space. However, among others in this field, the 20th century philosopher, Bertrand Russell, could not accept either of these views. Instead, he proposed a middle ground stand. According to him, there is an underlying reality beyond the human being's perceptive abilities – a real universe. But he contended that the essence of the real world is not simply as it appears to our senses. The concession is made to the idealist that since reactions are indeed happening at the place where we are, the conclusion of a three-dimensional space, according to our senses, could be illusory. Nevertheless, Russell contends that it is most likely not entirely illusory. He rather takes the role of the human's percepts to provide a link between the ends of a logically connected chain – the underlying reality that is there, independent of our perceptive reactions, and a fundamental rational description of this reality.

Thus, Russell does not contend that the human being directly 'sees' the real world. Rather, our percepts interact with the real world in such a way as to give 'hints' about the essential concepts that must underlie the universe. Since the mind is finite, (so that a human can never become omniscient!) we learn to use these hints successively to build up an understanding of the real world, though never expecting to reach the limit of complete comprehension. This is done using the ability to think rationally in order to infer from perceptions toward abstract underlying bases for natural phenomena.

In the way that our understanding is then developed, inferred conclusions should be expected to be wrong at least a part of the time. Indeed, the history of science reveals that this is so. The check on the validity of an idea in science comes from continually exploiting this idea in regard to its empirical predictions in physical experimentation. Sometimes ideas check in this way for many decades of experimentation, or even for centuries. But often, refined experiments lead to new observations that are not in agreement with the predictions of the ongoing theory. In these cases, some (or all) of the ideas that were previously inferred from the hints of the earlier perceptions and experimentation must be replaced with new ideas that can successfully explain the earlier theories' successful as well as unsuccessful results. In this way, then, the scientist and the philosopher hope to approximate closer and closer to the actual nature of the underlying reality that is the universe. This seems to me to be Russell's main stand in the philosophy of science.

One of the important aspects of the underlying reality that we wish to comprehend is the concept of space. According to his approach, Russell does not interpret the space of our percepts as the real physical space of the universe. He rather takes the view that the universe must be characterized by an underlying 'physical space'. This is a view that is in accord with the philosophic stand of realism. But it is 'abstract' rather than 'naive' realism.

With this abstract view, and the compromise that Russell takes between naive realism and idealism, what, in fact, are the features of 'space' that are hinted by our perceptions? One of these is continuity. The hint about this comes from the observation of motion, or from other observations in nature, such as the continuity in colour and form of a field of wheat. When I say that the points of space are 'abstractly' continuous, however, I have in mind a logical situation that is not directly perceivable, just as Euclid's axioms in geometry are assertions about relations between points and lines, as mathematical entities with a particular invented structure.

The abstract notion of continuity entails the following ideas: Given any three parameters, a, b, and c (from an infinite sequence of parameters), one may define relations between them called 'distance'. The sequence of points in space may then be defined as an imposed ordering of parameters, along with the concepts of 'further' and 'closer'. If the distance, $c - a$, is greater than the distance $b - a$, then c is said to be further from a than b is from a. Or, b is said to be closer to a than c is to a. The sequence of points then forms a continuum if it has the following property. If a and c are arbitrarily close together, there must always be room for the element b to be placed between them – i.e. b being closer to a than c is to a. Such a definition then implies that no matter how close two elements of this sequence may be, there must be an infinite sequence of other elements between them. Such a continuous space is called dense.

Since things are 'observed' to move in a continuous fashion, one has the hint that not only the space of the human being's perceptions, but also the abstract space that is used in the theoretical description of motion is continuous. From the physical point of view, this implies that no spatial point is inaccessible to moving matter.

Of course, the truth about Newton's absolute, abstract space is still in question. One of the first scholars to oppose it was Leibniz, who contended that one may specify a point of space only relative to other spatial points. That is, he did not feel that there is any justification in claiming the existence of an absolute space, with an absolute origin, as Newton did. Thus, Leibniz concluded that the physical (abstract) space should be regarded only in relative terms[8].

Question If our perceptions of three-dimensional space, such as our ability to point to where that chair is, are only hints about what is 'out there', then I see that it might be illusory that we are really in a three-dimensional space. If this is so, what would replace it?

Reply Your question presupposes that we, and all of matter, must be in some sort of space. But this is an interpretation of space that I have been saying is denied by the theory of relativity. According to this approach, 'space' is no more than the name for a set of ordered sequences of words, which one uses, in turn, to facilitate an expression of natural phenomena. This is not unlike the artist's use of paints and brushes to facilitate an expression of his particular impressions of the world. The emotion that is generated when one looks at Leonardo da Vinci's *Mona Lisa* – the reaction to the beauty in her face, the reality of life that one sees in her eyes, the joys, the sorrows, are not simply the sum of paints that Leonardo used on his canvas! I believe also that the particular reality portrayed in the painting is not uniquely expressible in this way. Perhaps the same set of qualities and truth might be portrayable in a piece of music – though not as many people would be able to respond in this case, as 'sounds' to human beings are more abstract a medium for the communication of emotion than visual means. In the same vein, should the *Mona Lisa* be painted by a 20th century Leonardo, in the form of 'modern art', even fewer human beings would respond to it, at least at first glance. Nevertheless, that latter more abstract forms of art may indeed be more accurate modes of expression, once one can understand the artists' languages!

Question Does this imply that as our intelligence may develop to higher planes in the evolutionary process, it may be possible for us to abandon the language of spatial coordinates altogether, and replace it with a more efficient abstract language to express the scientific laws?

Reply Indeed so. I would be willing to bet with high odds that as we reach the higher stages of evolution, we will evolve into stages of much more efficient languages to express scientific and philosophical concepts. Such language development will undoubtedly help in our progression toward acquiring further understanding of natural phenomena by providing us with better tools for the job of intellectual exploration. An extremely outstanding example of this sort in the history of science was Newton's invention of the calculus in order to facilitate a precise description of motion. It is salient that the concepts that underlie the idea of motion are physical, while the calculus is no more than a language to express these concepts. The concept of motion was a subject of study by the natural philosophers for millenia before Newton. But it took the more precise language of calculus to exploit these ideas fully, and then to be able to make precise mathematically formulated predictions of the outcomes of experiments, in order to check the validity of the underlying ideas.

On this point, it seems to me (in retrospect) that the scientific community was so thrilled with Newton's new scientific language and its power to predict that there were, unfortunately, many who began to equate physics, *per se*, with mathematics – sometimes playing with the equations at the expense of ignoring the physical notions and their further development, that the mathematical language was only to represent. (On this, it is interesting to recall that the popular press referred to the 'physicist' Einstein as a 'mathematician', in most of the press releases about his work.)

It seems to me that the identification of mathematics with physics, especially in the 20th century, was also a symptom of the impact of the philosophical stand of positivism. This approach started to take hold in the scientific community, with strength, in the early decades of this century. Then, one was instructed to look only for the most economic ways of representing the scientific data, and the logical relations between the elements of a language that serves to describe the observed facts. I think that with this point of view, science becomes totally descriptive rather than explanatory. It wasn't until Einstein's general relativity theory appeared that a serious attempt was made to break away from this view, and return to the notion of realism – the view that there is an underlying, universal reality, from which one can deduce particular physical consequences (among other particulars) that may be related to the reactions of the human being's instruments in physical experimentation.

Question   Can you predict anything about the character of a future language of science that would replace the present language based on spatial concepts?

Reply   If I had the slightest hint about this, I would be working on it! The future language of science may be a five-dimensional sequence of relations, or it may be some one-dimensional sequence of relations, or it may be a continuous, infinite dimensional sequence of relations, or it may be something that may not be expressible with the words, 'dimension' or 'sequence'. Frankly, I feel I am too far away from that stage of development in mathematics to be able to say anything intelligent about it, except for my faith that such progress will occur! I have faith that, in due course, another Newton will appear on the scientific scene and this new genius will be allowed by the scientific community to make the breakthroughs that we need in mathematics to express the scientific concepts that must come if we are to progress further in our comprehension of the universe.

To exemplify this state of affairs, I might mention a very important branch of mathematics that will become increasingly important for the development of the physics of the future: the subject of non-linear differential equations. Very little is actually known in this branch of mathematics and, unfortunately, it does not seem to be a popular field of inquiry among contemporary mathematicians. Still, I feel strongly that scholars in one field of study should not try to coerce the scholars of another field to work on problems that would be helpful to them. We can only hope that someday a large-scale effort and some brilliant insights might appear in the field of non-linear differential equations. This would most certainly help all branches of physics to make large strides forward.

Some of my colleagues in theoretical physics readily admit that the resolution of many of the ongoing problems in modern physics are in the solutions of non-linear equations, but they will stick with the linear equations because there are well-known, clear-cut methods of solving them. This reminds me of the story about the man who, one dark night, dropped his house key in the mud, in front of the house. When his friend, seeing him pacing beneath a lampost a block away, asked what he was doing there, he replied that he was searching for his key, which he admitted he lost in front of his house. When asked why he was searching there for the key, a block away from the house, he replied that it was dry and light under the lampost, while it is muddy and dark in front of the house!

NOTES

1 Concepts of space in the philosophy of science are varied. See, for example, B. Russell, *On the Philosophy of Science* (Bobs-Merrill, 1965); M. Jammer, *Concepts of Space* (Harvard, 1954).
2 The holistic view of the human body is taken, for example, in a therapeutic part of Chinese medicine called acupuncture.
3 See, for example, D. M. Armstrong, *Berkeley's Philosophical Writings* (Collier-Macmillan, 1965).
4 The original writings of Descartes on this problem may be found in: M. C. Beardsley, (editor) *The European Philosophers from Descartes to Nietzche* (Modern Library, 1960), p. 80.
5 B. Spinoza, *The Principles of Descartes*, (Open Court, 1905).
6 B. Spinoza, *Ethics* (Hafner, 1960).
7 For Russell's refutation of this view, see: Chapter 4, Note 2.
8 For Leibniz' relational view of space and time, see his *Monadology*, Supplementary Passage No. 2, Space and Time, in M. C. Beardsley, *ibid.*, p. 304.

7

Time

If a person should ask a neighbour what is meant by the word, 'time', the reply may be, in effect, that this is a parameter that characterizes duration – a change from the past to the present to the future. There are events, things, people, and so on, that were, but now are only a memory. Similarly, by using the method of induction, the human being's thinking action anticipates that there will be particular sorts of events to supersede those that are happening now.

As I write this chapter, at the very place where I now sit, about one hundred years ago, there was an uninhabited swampland, populated with the (now) rare birds and thick in foliage. Eighty years before that, British and American soldiers were firing rifle shots at each other, trying to gain this territory (in the war of 1812). At this same place, millions of years earlier, there were prehistoric creatures battling the elements, as well as each other, in their struggle for survival.

We 'know' about these things now from our studies of history and archeology, and from the information that has been passed on by word of mouth, from one generation to the next. But we must realize that all of this information had been pieced together from indirect and generally incomplete clues. Our 'knowledge' about the future, moreover, is based on an extrapolation from the present events and from what we think we know about the past. So, what we claim to know about the future is a speculative and inferential knowledge – including the claim that the sun will rise tomorrow morning!

For an understanding of 'time' we then have a situation similar to our previous conclusion about 'space' – there is a physical (abstract) time, and there is perceptual time. The latter time relates directly to the human consciousness, and the feeling we have of duration. The question then remains as to whether or not this perceptual time correlates in any way with a physical time – a 'time' that may exist to underlie a fundamental description of the real universe, in terms of objective laws of nature.

As with the philosophy of space, one may view the philosophy of time in terms of the extreme approaches of idealism, on the one hand, and naive realism on the other. The former would deny any reality in the idea of an abstract time, outside of the human being's perceptions, while the latter would assert that the human being's perceived time is the physical time that identifies with the real world – it is a feature of the real world that is directly projected onto the human consciousness. Then there is Bertrand Russell's intermediate view, that neither of these extreme positions is a valid philosophy of time. He would contend that it is rather that our perceptual reactions, that we interpret as time, serve as a link between the 'feeling' we have of duration, and the physically (non-tangible) real time.

As it was with his view of the concept of space, Newton believed in the existence of an absolute time. However, his contemporary, Leibniz, looked upon time as a relative concept – meaning that to specify a quantity of duration one may always refer to a conveniently located origin. For example, 'the time is 2 o'clock in New York' refers, in fact, to the location of the Earth's surface during its rotation about the north–south direction of its axis of rotation, relative to a simultaneous measurement of time at some other location on the Earth's surface, say at Greenwich, UK. Similarly 'it took three seconds for a ball to roll down a plane' is an assertion about the passage of time (correlated, for example, with the Earth's rotation) from some initial time measure to some final time measure. In the comparison with the Earth's rotation, the actual measure, '3 seconds', refers (with the standard that has been chosen in this example) to the Earth tracing out the fraction $1/28\,800$ of its daily night-day $360°$ rotation on its axis.

Newton did not deny this operational aspect of the time measure, and the way it is used in everyday applications. Nevertheless, he asserted further that all of the points of time, whose differences are related to time measures (of duration) refer in fact to a single, universal time axis, with an absolute origin – e.g. the time of the creation of the universe. Such an assertion about time implies that it would be meaningless to refer to any time before the creation of the universe, since 'time' only came into existence with the creation. This was the same view that was taken six centuries earlier, by Moses Maimonides, in the argument in his *Guide for the Perplexed* against Aristotle's conclusion of the eternality of the universe[1]. It is also similar to the view expressed another six centuries before Maimonides, by St. Augustine[2], in his *Confessions*.

Aristotle's conclusion, that Maimonides and Augustine rejected, was based on his definition of time as a measure of motion and the speculation that present motions must be created by earlier motions, so that the potential for motion has always existed and will always exist in the future, thus leading to the conclusion that time is eternal, with no beginning and no end!

There is also the mathematical use of the time parameter in physics to consider. As scientists, we study equations of motion. These are relations that we attempt to correlate with the results of experimentation. We then accept or reject the validity of the theory that the equations represent, as a bona fide law of nature. An equation of motion, in the classical sense, deals with the change in time of some variable property of a material system, such as the position in space of a material object. Thus, to describe motion, we introduce the parameter t into the appropriate equation. This parameter is then correlated with the observed motion of a material object, e.g. the motion of a block sliding down an inclined plane.

It seems that here we are using the time parameter in a way that would support the philosophic stand of naive realism. That is, we are correlating a perceived time measure with the parameter t, that appears in the expression of an objective law of nature – a parameter representing the abstract entity that has to do with 'physical time'.

Nevertheless, since the days of antiquity, to the Age of Reason of the post-Renaissance period, other views had been held. One example mentioned earlier, was Faraday's field concept, giving an adequate explanation for electricity and magnetism, according to the known facts about these phenomena in the 19th century. It is an essential point that in the field theory, time, as well as space, enters the description as a subdued parameter, not *directly* relating to any observable. This is in contrast with the appearance of the time parameter in Newtonian physics, in terms of the observed trajectory of a quantity of matter. That is, in the field theory, it is the set of appropriate field variables (mapped in an underlying space and time coordinate system) that play the role of the primitive variables, rather than the space and time points themselves. In the final analysis, comparison between theory and experiment must entail the use of numbers that are derived from the field variables, which in turn are connective relations, described as functions of the time parameter t – an independent variable – as well as the space parameters as independent variables. For example, the energy associated with the electromagnetic field is derived from an integration (a summation of continuously distributed variables) of a particular combination of the electromagnetic field variables, that has the property of being unchanged with respect to changes of the abstract time parameter of the formalism, i.e. the conservation of energy.

Another important use of the time concept in physical theory is in regard to its relation to another physical process, such as ageing. Here one makes a correlation between the physical process of irreversible human cell decay (for the ageing of a person), or the unwinding of the spring of a clock, and so on, and the change of an abstract time parameter. Even though one starts out by calibrating a scale of time measure, using a correlation with a physically evolving process, such as the unwinding of the spring of a clock, it is important to note that the statements made about the abstract time measure and statements about physical processes are different sorts of statements. They are not necessarily modes of expressing the same thing under all possible conditions of observation, even though a standard is invented in relation to a particular observational pattern. We will come back to this point later on, in our discussion of the so-called 'twin paradox'.

Since all physical observables do not correlate the human being's awareness of 'perceptual time' (duration) with 'physical time', it follows that in our theories of the physical universe, one cannot assert the view of naive realism in regard to the philosophy of time[3]. Neither does the view of idealism hold, as it is in conflict with the use of abstract time in field theory, such as the Faraday field theory of electromagnetism. It seems to me that, here also, Russell's intermediate view comes closer to a valid philosophy of time.

Question I am still confused about the different uses of the word 'time'. How do you distinguish between the 'time' that is involved in the classical cause–effect relations (cause coming earlier in time, and effect coming later), the 'time' in the definition of determinism in newtonian materialism, the 'time' involved in the human consciousness' becoming aware of duration, and the abstract time of field theory? Can you illustrate with reference to the human society?

Reply The roles of space and time in field theory, as abstract, non-observable parameters, that only serve the subjective purpose of facilitating a description of the world in terms of objective laws of nature, are quite different than the roles of space and time according to the implications of the philosophic view of materialism, that characterizes classical physics[4].

With the classical view, the separate material things, which are identified according to their locations in space and the times when they are there, are the elementary stuff from which the universe is built. The evolution of the predetermined motions of all material objects then relates to the time development of the universe, with this classical view.

On the other hand, the field approach seems to me to give a quite contrary interpretation of matter. For here there are no individual trajectories to be identified with separate material things – even though the view of separability of the classical approach forms a good approximation for the actual picture in field theory, under the appropriate conditions. But in fundamental terms, this view starts with the single closed system, rather than the collection of independent things (atomism). In the field theory, there is only one space and time language to be used in its description. But this language may be expressed in various ways depending on the frame of reference from which it is described. The idea, then, in seeking an understanding of the laws of nature as applied to a closed system, is to compare carefully the different features of the system, as expressed in as many different frames of reference as is imaginable, from the view of any one of them, then seeking the common elements (invariants) in the contents of the respective statements about the world that emerge. With this approach, the oneness of the system is somewhat akin to a single closed surface, like an inflated balloon, rather than a box full of different sorts of cookies. One can move a cookie about in the box, or remove some of them, without altering the basic characteristics of the system that is a box of cookies. But if, for example, a thumb is pushed into the surface of the balloon, tension is produced throughout the entire surface of the balloon!

If one should carry these ideas forward, as an analogue for a description of the human society, and attempt to ascertain which are the fundamental underlying elements – those associated with the atomistic, independent ego model, like the society of cookies in the box, or those similar to the abstract field theory of a closed society, like the surface of an inflated balloon, one would have to rely on the empirical facts about societies to verify or to refute one of these theories of society or the other – if the decision is to be based on a scientific approach to truth.

Of course, we have no 'established knowledge' at our present stage of knowledge of societies that will eventually be explicable in terms of the concepts of relativistic physics and its implied closed system – which would be my choice for a more valid model. Still, some of the experts on the problems of society do have hints from their studies of sociology, psychology and anthropology, that indeed reveal indications that this holistic view of a closed system model of society may turn out to be true after all[5].

A pertinent question that may be asked by the (perhaps naive) inquirer is the following. Suppose that a new-born human being should be placed in a society of kangaroos, in the outback of Australia. Imagine that the kangaroos should find it possible to accept this infant into their society and to take care of her needs, such as shelter, food, companionship and (a primitive sort of) love. As this baby grows, would she become a human being at maturity? Or would she be more like a kangaroo?

The atomistic view must conclude that no matter what sort of environment she grew up in, this person would remain a human being, even though she may not have learned the ways of civilization. That is, the assertion would be that this creature must remain human, strictly based on her genetic make-up.

The abstract field theory of the society would imply that at maturity this creature would be more like a kangaroo than a human being. Still, as an element of a closed system, the further implication must follow that the human being and the kangaroo are not totally distinct from each other. Rather, they are different manifestations of a single closed system, which is all of nature. At least, this is my interpretation of the philosophy that underlies the theory of relativity, when it is pursued to its logical extreme. Such a holistic view of nature – assuming that the universe is a system without actual separable parts – is the philosophic stand that was proposed by Spinoza, in the 17th century. Of course, our present understandings of the behaviour of a human being and the problems of society are at a very primitive stage, compared with our understanding in the physical sciences. These comments on the problem of society should then not be taken as more than speculative and conjectural, though induced by hints from observations of societies.

Question In your view, what are the basic differences that distinguish a human being from the other animate and the inanimate manifestations of the closed system that you are proposing as a model of the universe?

Reply The main difference, as far as I can see, is the human being's consciousness. It may be that a rabbit or a carrot, or even a rock, have some sort of consciousness, that is, that they can think in some way. But I do not have any scientific evidence that a rabbit has a consciousness that resembles that of a human in any way, nor that a carrot or a rock have any form of consciousness at all!

One of the important consequences of our having a consciousness is the feeling it gives us of duration, which we identify with time. It is this sort of time that I have referred to before as 'perceptual time'. This should not be confused with 'physical time' – that is expressed in terms of the language element (i.e. the parameter t) and is used in our equations of motion or in the field equations, to express laws of nature[6].

Of course, under special circumstances one may correlate these two types of time. Indeed, this is the reason that they are both called by the same name – 'time'. Still, one must keep in mind that these refer to entirely different concepts. The theory of relativity has taught the physicist to make this fundamental distinction. It was also emphasized in classical physics by Newton and in the contemporary period by Henri Bergson[7].

A second important feature of our consciousness is its ability to simulate an approximate detachment from our environment, and in so doing, acquiring an awareness of 'other'. This awareness is 'knowing'. It is important, however, that with this approach of the philosophy of relativity theory, approximate detachment can never mean complete detachment. This is because the human being, as well as all other components of the single closed system – the physical world – are only manifestations of this unified entity; they are not separable parts.

The view that I am expressing here is somewhat akin to the type of existentialism propounded by the contemporary philosopher, Martin Buber[8]. The fully closed system is summarized in his *I–Thou* relation. The state of approximate detachment is summarized in his *I–It* relation.

The idea of science, then, is to start with the complete universe – the universal existent to be associated with the physical world – and then to deduce observable manifestations of this single system, as derived particulars. In this context, it is important to note that one of the derived particulars from the underlying reality that is the universe is the human being's comprehension of it. That is, our understanding of the universe is one of its own manifestations! Since we are only finite beings, we can never become omniscient; we can never acquire a complete understanding of any manifestation of the universe, if such total understanding is infinite in extent[9]. But this is not to say that it is meaningless to assert a belief in the existence of the one universal that is the universe, with total underlying order. With this philosophy of science, which is Spinozist, it is our function, as scientists and philosophers, to probe continuously the single existent, of which we, ourselves, are basic components, successively approximating closer and closer to the 'truth' of this reality. But this is indeed an infinite number of steps, some forward and some (hopefully less) backward.

I believe that it is through this sort of intellectual activity that the human being transcends the crude materialism of naive realism, that our egos wish us to accept. Instead, we progress in such transcendence toward a more complete and realistic understanding of the world – a realism that is 'abstract' for the simple reason that our apparatuses to perceive and to measure are not the type that would directly respond to the objective truths of nature. But we have been given the capacity to think rationally. With this, I believe that we can capture at least a glimpse of the real world. At least, this is the view that I see taken by the philosophic stand of realism that underlies a full exploitation of the theory of relativity.

Some, who take the philosophic stand of logical positivism, deny that it is meaningful to talk about such 'underlying truths'. All that they would concede to exist are the logical relations between the numbers on the meters of our measuring instruments, that the human being is using to measure effects related to particular physical phenomena[10]. I do not believe that this is a valid view because it denies that we can attain any fundamental understanding of the world. It claims that all we can do is to describe it, but that we cannot explain it! I believe that this assertion is refuted by the history of science, which teaches that indeed we have attained some fundamental comprehension of the real world – meagre as it has been since the Stone Age, compared with all that there is to understand! This is exemplified in humankind's continual progress in the programme of philosophy and science, in gaining further understanding of the real world by means of rational and methodical comparisons of the hints received by our perceptive and instrumental probing devices and the logical implications of our hypotheses in physics.

NOTES

1 M. Maimonides, *The Guide of the Perplexed*, S. Pines, transl., (Chicago, 1963).
2 St. Augustine discusses the concept of time in his: *Confessions*, Book XI. For a translation see: A. Hyman and J. J. Walsh, (editors) *Philosophy of the Middle Ages* (Hackett Publishing Co., 1973), 2nd edn, Ch. 12–16, pp. 79–81.
3 A salient philosophy of time in terms of the measure of perceptable duration is in: H. Bergson, *Duration and Simultaneity* (Bobs-Merrill, 1965).
4 I have discussed a variety of meanings of the word 'time' in the French article: M. Sachs, Le concept de temps en physique et en cosmologie, *La Recherche* 9, 104 (1978).
5 For a discussion of the Gestalt theory, see: T. R. Miles, Gestalt Theory, *The*

Encyclopedia of Philosophy (edited by P. Edwards) (Macmillan-Collier, 1967), Volume 3, p. 318.

6 An interesting interpretation of 'time' from the point of view of the modern day existentialist philosophy is in: J.-P. Sartre, *Being and Nothingness*, H. E. Barnes, transl., (Philosophical Library, 1956), Part 2, Chapter 2.

7 See Note 3.

8 Martin Buber's philosophy is most clearly expounded in his book: M. Buber, *I and Thou* (Scribners, 1958).

9 In their exuberance about the successes of modern physics, some have claimed that the human race is indeed near the goal of all understanding of the universe. See, for example, S. W. Hawking, *A Brief History of Time* (Bantam, 1988). In his book, on p. 156, the author says: '. . . there are grounds for cautious optimism that we may now be near the end of the search for the ultimate laws of nature'.

This sort of optimism has occurred repeatedly in the history of science – usually at a time very close to a scientific revolution, when many of the believed ideas of where the truth lies were to be overturned shortly!

10 The epistemology of 'logical positivism' is clearly expounded in: A. J. Ayer, *Language, Truth and Logic* (Dover, 1952).

8

Space–time

Before the appearance of Einstein's revolution in physics, space and time were considered to be independent entities in the description of the physical world. That is, whatever was said about space – whether it was 'physical space' or 'perceptual space' – was taken to be wholly independent of whatever could be said about time. Thus, one was able to say that something happened at a particular place, without requiring any additional statement about when it happened. The latter information about the time of an event might be added (or not) to complete the description (or to leave it incomplete). The specific data about the 'where' and the 'when' were then assumed to be additive.

With the appearance of Einstein's relativity theory, the 'where' and the 'when' were no longer additive pieces of information, except in the approximation to the formal expression of the theory in which the classical (newtonian) description could be used for convenience. For in relativity theory, space and time become space–time – an inseparable unification into a four-dimensional system of coordinates.

In contrast with Newton's view, and in full agreement with that of Leibniz[1], space and time were now relativistic entities, with no possibility left to ascribe any sort of absoluteness to these concepts. Indeed, the concept of space–time became fully objective in that these four coordinates became the (minimum number of) continuously varying elements of a language that one observer or another may use to express the laws of matter. At the root of the theory is the assertion that the law of nature takes on the objective feature, in the sense that this is a relation about the behaviour of matter, independent of any particular frame of reference in which it is expressed, by any particular observer. This is Einstein's principle of relativity (or the principle of covariance). But in the maintenance of this objectivity, the space and time measures form a unified space–time measure relative to the reference frame in which it is expressed; hence the name – the theory of relativity.

The essential reason for the unification of space with time in the theory of relativity is a consequence of the imposed invariance of the forms of the laws of nature. To ensure that they will be in one-to-one correspondence in relatively moving frames of reference, it was discovered by Einstein that, when expressed with a language of space and time measures, what would be interpreted as a purely spatial (or a purely temporal) measure in one reference frame must necessarily be interpreted as a special combination of spatial and temporal measures in the other relatively moving frames.

Since the unification of space and time occurs by virtue of the relative motion of the different frames of reference in which the laws are described, by any one observer, and since motion is subjective (i.e. one can always find a coordinate frame that is in motion relative to any other one) it follows that to consider space and time as separate entities is strictly illusory. One must then consider them, within the context of the theory of relativity, as a fully unified set of coordinates. This is entirely analogous to the unification of the electric and the magnetic forces into a unified electromagnetic force, by virtue of the expression of their law objectively (i.e. independent of any relatively moving reference frame). That is, a purely electric force in one reference frame must be described as a unified electromagnetic force in other, relatively moving frames, if the law of electromagnetism is to maintain its objectivity, that is, if it is to be compatible with Einstein's principle of relativity.

Following from the necessary fusion of space and time into space–time, in order to satisfy the principle of relativity, it becomes a requirement of this theory to express the spatial and temporal measures in the same units. For example, one cannot add 2 apples and 3 trees to get 5 of anything, unless the apples and trees are reclassified into the same units, e.g. 'things of an orchard'. To express a time measure with the same units as the measure of distance one must then multiply the time units by a conversion factor with the dimension of distance per time – a number with the dimension of a speed. (One could equally convert all distance measures to have the same dimension as time measures by multiplying them with a conversion factor with the dimension of an inverse speed: time per distance. However, the former is the convention that is normally used.)

Thus, instead of a particular observer calling his time measure t seconds, he calls it ct cm, where the conversion factor c is a universal speed. (At this stage of the argument, we do not yet know that this turns out to be the speed of light in a vacuum). What is taken by an observer to be a spatial measure of x cm in one frame of reference must then be expressed by him as the combination $(ax' + b(ct'))$ in a different frame of reference that is in motion relative to him. The dimensionless coefficients in this transformation, a and b, must describe the relative motion: they are dependent on v/c – the relative speed of the moving frame divided by the speed of light.

Thus, the observer refers to the time measure in the moving frame (relative to himself) as ct' cm. In a third frame of reference he would refer to it as ct'' cm, and so on. A salient point is that all relatively moving observers must agree on the way of expressing their respective time measures

with a spatial dimension – they must all agree on the conversion constant
c. If this were not so, then one would not be able to separate the time
component from the space components in any arbitrarily expressed
space–time measure. But there are physically related empirical reasons, for
example the expressions of the laws of conservation (of energy, momentum
etc.), that require such a separation.

The law of conservation, such as energy conservation, states that a
particular quality of a material system, called 'energy' – the quantitative
ability of the matter to do work – is constant for all values of time. That
is to say, various types of energy may be converted into one another –
kinetic energy, chemical energy, potential energy – but according to this
law of conservation, the total energy content, of all sorts, must be the same
to any observer, whenever he might measure it, according to the relative
time measure of his reference frame. Now if a conservation law is indeed
a law of nature, then according to the principle of relativity, it must have
a corresponding form in all other relatively moving reference frames. If,
for example, energy is constant with respect to the time measure ct in one
frame of reference, there must be a quality called 'energy' that satisfies a
law of energy conservation, in a different relatively moving frame, which
would be constant with respect to the time measure, ct' of that frame. To
know that the respective energy conservation laws are indeed in one-to-one
correspondence in the relatively moving frames of reference, it is then
necessary to translate the languages of the respective reference frames by
applying the correct transformations in which the ct of one reference frame
becomes the correct combination (ct', x'), so as to preserve the forms of
the laws of nature in these different coordinate systems. When the motion
is constant, rectilinear relative speed, the correct translations of the space
and time measures are the Lorentz transformations, as first discovered in
the case of Maxwell's equations for electromagnetism. The latter special
case, in which the relative motion of the reference frames in which the
laws are compared and required to be in one-to-one correspondence, is
called the theory of special relativity.

Thus, to be able to separate the time measure from all relatively moving
frames of reference, a given observer must express each of them as: ct, ct',
ct'', ... cm. The full unification of the space and time into space–time then
necessitates the appearance of a universal speed – that is, a speed, c, that
is the same when expressed by any observer, referring to its value in any
relatively moving reference frame[2].

To determine the magnitude of this speed, it is necessary to perform an
experiment that would check the validity of a law of nature – any law of
nature! As we have discussed previously, the first law that was used to
identify this speed was the electromagnetic theory, as expressed in terms
of Maxwell's field equations. From this particular check of the correspondence
of the theory with experiment, it was discovered that the required conversion
factor – the speed c – is the speed of light in a vacuum, which in turn is

the speed of propagation of the electromagnetic force in a vacuum. Once the numerical value of this speed was found, the theory of relativity then interpreted c as the maximum speed of propagation of any force, in addition to the electromagnetic force.

With this argumentation, I do not believe that it is necessary to postulate the universality of the speed of light as a separate axiom for relativity theory, as originally done when Einstein proposed his theory of special relativity, in 1905. As we have seen, this follows logically from the principle of relativity when this principle refers to the laws of nature in terms of the fused space–time language.

Question Does the unification of space with time imply that they can be treated on the same footing? Does this imply the possibility of time travel, because we can travel in space?

Reply Your question refers to a quite common misconception among the lay public about the meaning of space and time in the theory of relativity. The answer is that the theory of relativity does not, in itself, imply the possibility of time travel! The unification of space and time into space–time in this theory does not mean anything more than a necessary enrichment of the language that must be used to express the laws of nature, so that they remain objective (independent of the reference frame). It does not imply that one may travel from one point to another in time, as one travels between spatial points.

There have been many science fiction stories woven about this misconception, mainly because of the confusion on the part of the authors (if they take this idea of time-travel seriously) between 'experienced time' on the one hand, and the abstract 'physical time', expressed with the independent variable t, on the other, in the mathematical expressions of the laws of nature. Of course, we do have experiences of 'seeing' the past. When one looks into the night sky, he or she is observing many stars the way they were when they emitted the light, many years in the past; it has taken millions of years for the light from some stars to reach us. Indeed, at the time of observation of a particular star in a far away galaxy in the night sky, this star may no longer exist, having burned itself out millions of years earlier!

But there is generally no implication in relativity theory about time-travel into the past or future. Such travel would be a physical experience – a sequence of physical events. The time parameter, on the other hand, is not more than a sequence of 'words' that an observer invents for the purpose of expressing a physical law.

Of course, if there should exist a law of nature relating to the experience of time-travel, then, according to the theory of relativity, such a law (as any other law of nature) should be expressible with a language that involves the spatial coordinates as well as the time coordinates in terms of a unified space–time. This must be such that the law would be in conformity with the principle of relativity, as any other law would be, according to this theory. But the theory *per se* does not predict time-travel. In any case, such physical experiences that involve time-travel have never been detected, to this date.

NOTES

1 G. W. Leibniz, Space and Time, in his 'monadology'. See: M. C. Beardsley, (editor) *European Philosophers from Descartes to Nietzsche* (Modern Library, 1960), p. 304.
2 H. Minkowski, 'Space and Time', in: A. Einstein, H. A. Lorentz, H. Weyl and H. Minkowski, *The Principle of Relativity* (Dover, 1923), p. 75.

9

The principle of relativity – from Galileo to Einstein

The principle of relativity asserts that if a set of relations that underlie a particular physical phenomenon is a 'law of nature', then it must have the same form in all possible frames of reference, relative to any particular observer.

My use of the word 'observer' in this context is not intended to have any anthropomorphic denotation. It refers only to the frame of reference of an interacting component of a closed system, from which the laws of nature are described. This may be the reference frame of a human being, or a beetle on a planet in a distant galaxy, or it may be the reference frame of a proton in rapid motion within the boundaries of an atomic nucleus, or that of a star within a particular galaxy in the outermost regions of the universe.

With this meaning of 'observer' in mind, the principle of relativity implies that an observer may respond to certain physical phenomena according to corresponding laws of nature for these effects. Other observers who may be in motion relative to the first (or at relatively displaced spatial and/or temporal positions, or they may be Martians, thinking in different concepts and speaking entirely different languages!) respond to the same phenomena from the frames of reference that most naturally describe their respective places in the universe. The principle of relativity then requires that when the given observer learns to translate their languages, each one of their expressions of a law must be in one-to-one correspondence with his original expression for this law and the expressions for the same law in all other possible frames of reference. That is to say, the 'law of nature' is asserted to have objective connotation.

Of course, in comparing the frames of reference of a human being and a dolphin, there may be some difficulty in communicating to the point of knowing whether or not there is agreement on a particular law of nature, in order to check its validity. But this does not mean that such a one-to-one correspondence between the human being's and the dolphin's comprehensions of the natural law in question does not exist!

In the case of relatively moving frames of reference with observers in them who do communicate, one may check the validity of the principle of relativity as soon as the correct transformations between the 'languages' of the respective observers are known. This was realized by Galileo, who was the first to recognize the scientific truth in the assertion of this principle. He discussed it in his *Dialogues Concerning the Two Chief World Systems*, where he emphasized that the laws of motion should be independent of where one wishes to place the origin of the spatial coordinate system, for convenience – should it be at the centre of the Sun or at the centre of the Earth, or anywhere else in the space of the universe!

Thus Galileo upheld and defended the Copernican view that the Earth 'moves', as do all of the heavenly bodies. But he went beyond Copernicus by asserting that, because of the relativity of motion, neither is the Sun at an absolute centre of all motions of the heavenly bodies, because there is no absolute centre of space in the first place! This idea has come to be known as 'Galileo's principle of relativity'. It implies that it is as true to say that the Earth 'moves' about the Sun as it is to say that the Sun 'moves' about the Earth – so long as the law of their relative motion has a form that is independent of where one wishes to place the centre of the spatial coordinate system.

In arriving at his principle of relativity, Galileo reasoned as follows. Consider a quantity of matter to be at a spatial location x in some coordinate frame. Let us say that this is Sally's location, standing on the deck, relative to the aft tip of a ship. When this ship moves in a straight line path, with constant speed v, relative to a stationary observer – Bill, who is standing on the dock – then Bill should see that the ship is changing its position relative to him by vt cm in t seconds. The position of Sally, on the ship, as viewed by her friend Bill, on the dock, would in this time be shifted from the location x to the location $x' = x + vt$. Still, as far as Sally is concerned, she claims to be x cm from the aft end of the ship. If she would not look away from the ship, Sally would not be aware that the ship is moving away from Bill's position on the dock. Her language for the spatial measure on the ship would still be x. This language would coincide with Bill's language, x', only if there would be no relative motion between them (i.e. if $v = 0$).

Had Sally looked at her friend, Bill, on the dock, as she was in motion relative to him, she would have claimed that his position would have changed (in the same time that Bill considered) by $x = x' - vt'$ cm. Bill's statement and Sally's statement are equivalent if one can transform from one to the other (by transposing from one side of Sally's (or Bill's) equation to the other, thereby generating Bill's (or Sally's) statement. This is indeed the case above, provided that we set $t' = t$. This means that physical events look the same to Sally and Bill, according to their respective coordinate frames of relative spatial measures and a common time measure. That is, Galileo tacitly assumed that there was an absolute time measure while the spatial measures were relative, as indicated above, in the expression of a law of motion.

Galileo then asserted that the laws of nature should be the same to either of the observers, Sally or Bill (or should we say, Maria or Giuseppi), who are in relative motion at a constant rectilinear speed (this is called 'uniform motion'). To check if this principle of relativity is correct, one merely has to substitute into Bill's expression of a natural law, in terms of the space and time words (x', t'), the following transformations from Sally's space and time words:

$$x' = x + vt \quad \text{and} \quad t' = t$$

According to Galileo's principle of relativity, one should then arrive at the same expression of the law of motion, though in terms of the words (x, t) rather than (x', t').

Suppose, for example, that Maria should throw a ball downward from the Leaning Tower of Pisa, with an initial velocity of v_0 cm/sec. Galileo discovered that the law of motion that governs the distance fallen in t seconds is

$$x = v_0 t + \tfrac{1}{2} g t^2 \tag{1}$$

where g is the constant acceleration of free fall, due to gravity. Suppose now that as the ball is descending toward the ground, Giuseppi ascends upwards in a balloon, at the constant speed of $-v$ cm/sec. (The minus sign means that he is moving in a direction opposite from that of the falling ball.)

If Maria's expression (1) for the law governing the fall of the ball is correct, then according to Galileo's principle of relativity, Giuseppi should deduce from his observation of the falling ball that the law of nature that governs the motion is

$$x' = v_0' t' + \tfrac{1}{2} g t'^2 \tag{2}$$

Since, according to Galileo's rules, time is absolute $(t' = t)$, the latter equation takes the form:

$$x' = v_0' t + \tfrac{1}{2} g t^2$$

Giuseppi ascends at the rate of $-v$ cm/sec, from Maria's view. But from his view on the balloon, as the stationary observer, he would see the tower descend at $+v$ cm/sec, along with the ball that was thrown downward at $+v_0$ cm/sec. Thus, Giuseppi should see the ball start its descent from the top of the tower at the speed $v_0' = (v_0 + v)$ cm/sec. Substituting this equation, and Galileo's transformation $x' = x + vt$ into (2), we see that this is the same as (1).

This recipe seemed to work in regard to the motions governed by the laws of gravitation. This was even after Newton had discovered the Law of

Universal Gravitation, when it was found that, in contrast with Galileo's law, the acceleration of the body due to gravity, g, is not a constant. Rather, it is a function of the distance between the gravitationally interacting bodies, such as the distance between the falling ball and the centre of the Earth, in the example above.

Two and a half centuries after Galileo's discoveries, it was found that if one should insist on the universality of Galileo's transformations between the space and time 'words' of the relatively moving reference frames, then Galileo's transformations $x \rightarrow x'$, do not work in regard to the laws of electromagnetism, as expressed with Maxwell's field equations. Instead, it was found that these field laws would be in one-to-one correspondence in different frames of reference that are in relative uniform motion only if Galileo's rules for space and time transformations would be altered to the following form (again assuming a one-dimensional space, for simplicity):

$$x' = \frac{x + vt}{\left(1 - \left(\frac{v}{c}\right)^2\right)^{\frac{1}{2}}}$$

$$t' = \frac{\left(t + \frac{vx}{c^2}\right)}{\left(1 - \left(\frac{v}{c}\right)^2\right)^{\frac{1}{2}}}$$

where c is the universal speed discussed previously, numerically equal to the speed of light in a vacuum. These are the 'Lorentz transformations'. They underlie the covariance (i.e. the objectivity) of the laws of nature in special relativity theory.

Lorentz originally considered these transformations as a physical action of the aether (that he assumed to be the material medium to conduct light propagation) on the instruments that measure lengths and times. In particular, he tried to interpret the negative effect of the Michelson–Morley experiment (in which an interferometer was used to measure the differences in speeds of light in different directions in the aether, relative to the Earth's motion) in terms of the contraction of the arm of the interferometer that was oriented in one direction, different than the contraction in the perpendicular direction. But Einstein rejected this interpretation of the Lorentz transformations as a physical effect on measuring instruments. Rather, he interpreted them as 'translations' of subjective language parameters, as scale changes, as we have discussed above, in the expressions of objective laws of nature. He then asserted that the aether is indeed a superfluous concept in our understanding of light propagation. Thus, Einstein maintained the spirit of Galileo's principle of relativity, except that he found that the correct transformations of x and t must be those of Lorentz, rather than Galileo. But, in his view, the Lorentz transformations have nothing to do with any physical interaction between a conducting medium and measuring instruments.

A salient point is that, according to the principle of relativity, the space and time transformations are a way of translating the 'words' that are used to express a law of nature – they are not a solution of a particular law of nature that one may be talking about. It then follows that either Galileo's rules for transforming coordinates are correct for all phenomena, or Einstein's rules are the correct ones for all phenomena – they cannot both be correct! Thus, if we are to accept Einstein's rules for space and time transformations as the correct ones, they must also apply to the laws of motion that Galileo was considering.

It is clear that the Lorentz transformations do apply to the experimental conditions that Galileo was investigating, such as a block sliding down an inclined plane, or even the motion of a planet in its orbit, relative to the Sun's position. This is because the relative speeds, v, in his experiments, were very much smaller than the speed of light, c. Thus, the ratio v/c is close to zero, in these applications. Taking account of the accuracy of measurement in Galileo's experiments, it would not have made any difference in his calculations if he had used the Lorentz transformations, for the ratio v/c could be approximated by zero, in which case these transformations reduce to his own transformations!

It is to be noted that with $v/c = 0$, one has a formula that would correspond to the description of 'action-at-a-distance', where one has an infinite speed of propagation of forces between interacting matter. This limit then corresponds to Newton's assumption in his physics, that in turn was built on Galileo's considerations. But when c is a finite number, this corresponds, physically, to a finite speed of propagation of forces between interacting matter – leading, for example, to the notion of gravitational radiation, rather than Newton's model of gravitational forces that act at a distance, spontaneously.

Thus, it is important to note that so long as c is a finite number, the transformations of Galileo and those of Lorentz are not exactly the same. In the cases where material particles move close to the speed of light, (as in high energy physics experimentation of the present period in physics) the Galilean transformations are a very bad approximation. One is then forced to use the form of Einstein's special relativity theory equations (i.e. to use the unapproximated Lorentz transformations). In the extreme limit, where v/c is equal to unity, corresponding, e.g. to the propagation of light itself, there is no longer any semblance of a classical description.

Recall that, thus far, we have been discussing special relativity – the special case where the relative motion between frames of reference is a constant speed in a straight line – the so-called 'inertial frames of reference'. But the theory of relativity does not generally require this special type of relative motion in defining the objectivity (covariance) of the laws of nature. When the motion is arbitrary, such as accelerated frames of reference relative to the stationary frame of an observer, then we extend the formal expression of the theory to 'general relativity theory'. In the latter case, the Lorentz transformations are no longer valid! They must be replaced with a more general form for the space–time transformations between relatively moving frames of reference. General relativity will be discussed in more detail in Chapter 18.

Question   What was really revolutionary about Einstein's theory of relativity, if it was only an extension of Galileo's principle of relativity, incorporating the time parameter with the relativity of the spatial parameters of the classical requirement?

Reply   The main conceptual change came with Einstein's reinterpretation of the space and time coordinates as not more than subjective language parameters to express objective laws of nature. In classical physics, space and time are more than a language – they are locations within an absolute continuum of points. It is like an empty vat that is there, whether or not there is anything in it. As I will try to indicate in Chapter 18 when Einstein's theory of relativity is fully exploited, one no longer says that matter is in space and time. It is rather that the relations between the space–time points – their geometry and algebra – is not more than a logic that is to underlie the language expression for the laws of nature, analogous to the syntax of ordinary verbal language.

Question   How might the principle of relativity apply to the human society?

Reply   One must not think for a moment that we are even remotely near to discovering a set of explicit field equations that underlie the behaviour of human beings, nor are we aware that such a precise formulation in mathematical terms could even exist. However, if one should assume that the principle of relativity generally applies to the human society, which I, perhaps naively, believe to be the case, then this principle does imply some interesting consequences. It implies, firstly, the existence of invariant relations between human beings, that is, relations that are independent of the particular frames of reference in which they may be expressed. These may be reference frames of national classification (American or Russian or Thai or Israeli, etc.) or frames of reference according to worker status (blue collar or white collar or professional or proletariat, or bourgeousie, etc.) or according to economic philosophy (communist or socialist or capitalist, etc.) or frames of reference according to religion (Christian or Jewish or Muslim, or Hindu, etc.) or any other subclassification of human beings.

In these cases, instead of labelling frames of reference in terms of space and time coordinates as it is done in classical physics, to locate inanimate things, we have referred to the frames of reference in terms of national labels, or labels that signify ethnic-cultural characteristics, or worker status, etc. My contention is that all of these, though real distinctions between possible frames of reference, are not essential distinctions to underlie the fundamental nature of the human society. I think that the essential distinctions are rather the basic drives that motivate the mutual actions of all people – not as separable parts of a society, but rather as 'modes' – each reflecting the entire society, as one closed entity. This view follows the pattern of the essential aspects of a material system according to the philosophy that underlies the theory of relativity, with the mutual interactions viewed as the component elements of the closed system, rather than viewing it as an open set of separable atomic constituents.

Because, as human beings, we have seemingly independent consciousnesses that can enter a state of very weak coupling with the rest of the closed society, we choose (in a 'predetermined fashion', according to this philosophy) to behave in one way or another – which may be for the well-being of the society, or sometimes (primarily) for what we feel to be in the best interests of the 'self'. But whichever the behavioural pattern, within societies or between different societies, it can be largely a function of education to inform the individual of his or her choices of actions and their consequences. Education should lead to the understanding, in accordance with the principle of relativity, that as human beings, we must seek out the invariants of social behaviour. These are the features that truly underlie the nature of the real world, of which human beings are not more than particular manifestations – just as a ripple on a pond is really not more than one inseparable manifestation of one pond. Just as the ripple cannot be removed from the pond, and then studied on its own as an independent entity, so a human being cannot be separated from society and understood as an independent entity, apart from society. I believe that to understand this philosophy could lead toward universal mutual respect, humanism and peace.

10

Violations of 'common sense' notions of distance and simultaneity

As we discussed previously, according to special relativity theory the correct translation of the space and time 'words' from the language of one reference frame to another, in order to maintain the forms of the laws of nature (the principle of relativity) in both frames, where one frame travels at a constant speed v in a straight line relative to the other, was found to be given by the Lorentz transformations. When, for example, the relative velocity between the two frames of reference is in the x-direction in 3-dimensional space, with the x and x' axes (of the respective frames) parallel, then the laws of nature will maintain their forms under the coordinate changes:

$$x' = \frac{x + vt}{\left(1 - \left(\frac{v}{c}\right)^2\right)^{\frac{1}{2}}}$$

$$y' = y$$

$$z' = z$$

$$t' = \frac{t + \frac{vx}{c^2}}{\left(1 - \left(\frac{v}{2}\right)^2\right)^{\frac{1}{2}}}$$

At this point it is important to keep in mind the meaning of the symbols (x, t), (x', t'), (x'', t''), ... as relative measures, in the sense of using different calibrations in the relatively moving frames of reference. These coordinate notations do not at all refer to physical properties other than their use in facilitating a description of the latter. In the Lorentz transformation formulas above, x' stands for an abstract spatial coordinate used by one observer, in her own frame of reference, relative to some convenient point in that reference frame, such as her telescope at the site of a rocket launching pad. On the other hand, x is a spatial coordinate determined by the same observer but in a different frame of reference, for example the rocket ship – as the observer watches it travel away from her at the speed v. Similarly, t' stands for the observer's time measure, calibrated with reference to some physical clock in her own frame, say her wrist watch, while t stands for her measure of the time passing in the moving frame of reference (the rocket ship). In classical physics, the latter two time measures would be the same; in relativity physics they are not, $t \neq t'$. The coordinates (x, t) and (x', t') of the relatively moving reference frames (the launch pad and the rocket) are both defined from the given observer's point of reference – whichever frame is chosen for this is strictly arbitrary. That is to say, we assumed above that the observer (on the launch pad) is the stationary one. But we could have equally represented the frame of the rocket as stationary, with the launch pad moving relative to it (backwards) at $-v$ cm/sec, in all of the transformation formulae above. According to the principle of relativity, such an alteration should not change the forms of the laws of nature – be they expressed from the reference frame of either the launch pad or the rocket.

With the alteration from the classical theory whereby one replaces the galilean transformations with the Lorentz transformations, there seem to be several predictions which appear to defy 'common sense'. The first concerns the notion of simultaneity. For since the time measure is not absolute ($t \neq t'$), it follows that if, in a particular frame of reference, two events are seen to happen simultaneously, at a particular point in time, they would not appear to happen simultaneously from a different frame of reference that is in motion relative to the first. That is, the concept of simultaneity is only relative to a particular reference frame.

This result can be seen with the aid of the Lorentz transformation formulae. Suppose that one event happens at the place x_1 at the time t_1, while a different event happens at (x_2, t_2). In relatively moving frames of reference, the corresponding space and time measures are the Lorentz transformations above, inserting the subscripts '1' and '2' for the separate events. It then follows that a time difference in one reference frame relates to a time difference in the other as follows:

$$(t_2' - t_1') = \frac{(t_2 - t_1) + \left(\dfrac{v}{c^2}\right)(x_2 - x_1)}{\left[1 - \left(\dfrac{v}{c}\right)^2\right]^{\frac{1}{2}}}$$

Thus, if two events in the unprimed reference frame are simultaneous, i.e. if $t_2 = t_1$, they would not be simultaneous in the primed reference frame, i.e. $t_2' \neq t_1'$. In this case,

$$t_2' = t_1' + \frac{\left(\dfrac{v}{c^2}\right)(x_2 - x_1)}{\left[1 - \left(\dfrac{v}{c}\right)^2\right]^{\frac{1}{2}}}$$

It also follows that, according to this formula, the magnitude of t_2' could be greater (later) than t_1' or it could be less (earlier) – depending on whether x_2 would be greater or less than x_1. Thus the concepts of 'earlier' and 'later' also become terms that are relative to the observer's frame of reference.

According to the theory of relativity, then, if two balls should be propelled in opposite directions at the same time, each at the same speed toward equidistant walls of a squash court, an observer at rest in the middle of the court should ascertain that they would hit the opposite walls simultaneously. However, someone who flies by in a rocket ship, observing this game, would conclude that the balls must hit the opposite walls at different times. Then who is right?

According to my view of the philosophical basis of the theory of relativity, the answer lies in the fact that there is not, generally, a one-to-one correlation between the parameter that is used for a time measure, on the one hand, and a physical interaction, on the other, such as 'ball hits wall'. The correlation only exists when one calibrates his clock so as to coincide with some physically evolving process, say in the frame of reference of the interacting objects, which in this example is the squash court. That is to say, if one should be observing phenomena in the squash court, from a frame of reference that is in motion relative to it, he must first identify the part of his time measure that has to do with his relative motion, i.e. relative to the interacting ball and wall. When he would do this, using the Lorentz transformation, he would then agree with the observer who is stationary in the squash court that the collisions of the balls with the opposite walls actually happened simultaneously. This is in spite of the fact that, according to the moving observer's language of space and time measures, it would not appear at first glance that they ought to be simultaneous events!

Even though this conclusion does not agree with the conclusions of most present-day relativists, it seems to me to be a logical consequence of the theory of relativity, if one should be willing to assert that a real universe exists whose essential physical features are independent of the states of motion of observers who happen to be probing this real world, in one way or another. A denial of this philosophic stand (realism) like the stand of idealism, has been defended by many past scholars, such as George Berkeley[1]. But I am starting this analysis of the theory of relativity with the assumption that there is a real world that we can comprehend to some degree, a world whose fundamental features are independent of whether or not a human being happens to be probing it.

I believe that Einstein took this stand in his development of the theory of relativity, especially so after 1915 when he introduced the generalization from the earlier investigation of the theory of special relativity – which did seem to rely on a philosophic stand of operationalism – to general relativity – which is based on the abstract field concept, with an interpretation in terms of realism. With the latter view, while the space and time measures refer to relative language elements to be used in the expression of the laws of nature, the laws themselves, that govern physical interactions, are indeed taken to be objective with respect to changes between their expressions in the different reference frames (according to the principle of relativity).

A second apparent violation of 'common sense' notions is the 'Fitzgerald–Lorentz contraction'. Taking what has been said so far about space and time measures, it follows that if x_2' and x_1' are the simultaneously measured ends of a measuring stick in, say, Sally's frame of reference then if x_2 and x_1 should be the simultaneously measured ends of the stick in another frame of reference (in motion at the speed v in the direction of the stick's length, relative to Sally's reference frame) then the relation of the coordinates for the spatial separations of the ends of the stick should be

$$L' = \frac{L}{\left[1 - \left(\frac{v}{c}\right)^2\right]^{\frac{1}{2}}}$$

where $L \equiv (x_2 - x_1)$ and $L' = (x_2' - x_1')$.

It seems, then, that with the inverse relation,

$$L = L'\left[1 - \left(\frac{v}{c}\right)^2\right]^{\frac{1}{2}}$$

a stationary observer of a moving stick would claim that its length, L, is shorter than its length L' in the reference frame of its own coordinates, i.e. the reference frame of an observer sitting on the stick. This is called the Fitzgerald–Lorentz contraction.

To exemplify this conclusion, suppose that Sally is capable of propelling a metre stick on the ice at one half of the speed of light. It is made to move in the direction of its length. Sally would then say that, according to her observation, its true length is equal to

$$(100 \text{ cm}) \times \left[1 - \left(\frac{\frac{1}{2}c}{c}\right)^2\right] \simeq 85 \text{ cm}$$

Suppose now that there is a hole in the ice that is 95 cm in diameter. When the stick passes over this hole, would it fall in? Sally might conclude that since the hole is stationary (in her reference frame) and the stick is shorter than its diameter by 10 cm, the stick must fall into the hole when it passes over it.

On the other hand, consider a bug, B, sitting on the stick on the ice. It would watch the hole approaching at $-c/2$ cm/sec. B must then claim, if it interprets its data just as Sally did, that while the stick on which it sits is 100 cm long, the diameter of the hole is only

$$(95 \text{ cm}) \times \left[1 - \left(\frac{-\frac{1}{2}c}{c} \right)^2 \right] \approx 81 \text{ cm}$$

That is, the bug would conclude that the diameter of the hole is smaller than the length of the stick, and therefore the stick would not fall into the hole.

According to the theory of relativity, the nature of the physical effect cannot depend on the frame of reference in which this effect is described. But Sally and the bug came to opposite conclusions about whether or not the stick would fall into the hole in the ice! If they are both correct, then this would be a logical paradox. If the theory is to be conceptually valid, it must be logically consistent. Thus, either Sally or the bug must be telling the truth, but not both of them! Then who is right?

The answer is that, according to the definition of the space and time measures in relativity theory as language parameters that are not directly physical entities, it must follow that the bug is right in its conclusion that the stick will not fall into the hole. Independent of anybody's view of this situation, the stick will not fall into the hole for the physical reason that it is 100 cm long and the hole is only 95 cm in diameter. But the bug's reasoning in coming to this correct conclusion was not valid. The trouble is that neither Sally nor the bug had yet learned about the theory of relativity. Each of their conclusions was based on their respective deductions about relative lengths, as determined from their subjective observations of the stick and the hole in the ice.

If Sally had known about the theory of relativity, she would have been able to conclude that her deduction about the length of the stick was fallacious – because of the fact that the stick was in motion relative to her (at the speed $c/2$), and she didn't take this into account. Applying the Lorentz transformation, she would have calculated the stick's proper length to be

$$\frac{85 \text{ cm}}{\left[1 - \left(\frac{\frac{1}{2}c}{c} \right)^2 \right]^{\frac{1}{2}}} = 100 \text{ cm}$$

Since the hole in the ice is stationary with respect to Sally, she would know that its proper diameter is only 95 cm, as observed. Thus she would have concluded that the stick should not fall into the hole in the ice.

The bug's conclusion about this was correct, but in making its calculations, had it known about the theory of relativity, it would have found with the Lorentz transformation formula that the hole diameter is 95 cm, now in agreement with Sally's conclusion, rather than the 81 cm, as claimed before.

Question I recall reading in a popularized account of relativity theory that lengths and times are, in its view, operational, in any frame of reference. Then what do you mean by the proper (or real) length of a stick, as distinguished from its length that would be determined in any other frame of reference, when it would not be called 'real'. Does not your reference to the special feature of the 'proper space frame' of the stick revert to the idea of an absolute space?

Reply As I have indicated before, the earliest expressions of the theory of relativity were indeed strongly influenced by the positivistic stand, whereby lengths and times were treated only in operational terms. But as Einstein extended his theory, both mathematically and logically, he was led in a natural way to a re-interpretation of the theory – in which the positivistic approach was replaced by an approach of abstract realism. This is the view that there is a real, substantive universe that one may talk about, whose essence is matter and whose manifestations are expressed most fundamentally in terms of the mutuality of a closed system, rather than in terms of spatial and temporal concepts.

When I spoke of the 'proper length' of a stick, I referred to the language element associated with a measure of length in which the measure is calibrated in the frame of reference of the system of molecules whose mutual interactions hold each others' centres in some stable configuration. If no external pressures are applied to the real stick, then it should not change its equilibrium spatial configuration – that is, it should not shorten, physically, because an observer happens to see it in motion relative to himself! Thus, the use of the term 'proper frame of reference' does not at all imply a reversion to the concept of an absolute space. It is rather an expression of the special role that interaction plays in the theory of relativity[2]. There is still no absolute space within the conceptual structure of this theory – nor, for that matter, is there any sort of space, as a thing-in-itself.

Most contemporary physicists, including those who call themselves relativists, claim that the moving stick shortens, physically, in the example discussed above. But they are ignoring the logical paradox that ensues from this interpretation – that it really shortens to one observer of the stick but it does not really shorten to another observer of the same stick (the one who rides with it). Thus it must be concluded with this interpretation that the stick both shortens and it does not shorten!

I think that most people's reluctance to accept the idea that the stick might not really shorten because of its relative motion is the difficulty in accepting the idea that space (as well as time) are purely relative language constructs, which are invented for the purpose of facilitating an expression of the laws of physical effects – probably for want of a better language! Whether or not most physicists will admit this, it seems to me that what they are, in fact, holding onto is the earlier newtonian view[3] of absolute space, even though they now concede that this absolute space can distort under appropriate conditions of relative motion. But this view is diametrically opposed to the meaning of space in relativity theory, as not more than a language construct – not something physical in itself. In this view, there is no objective space that can do anything physically, such as deforming, because it is not a physical entity. It is rather that it is only the scale of a language expressed in a particular frame of reference that must be altered, in order to describe correctly objective physical laws in relatively moving frames of reference. The idea of an absolute physical space that can deform is, in contrast, much closer to Lorentz' original interpretation of his transformations, as a physical effect of the aether on the substance of matter in motion. In Lorentz' model, the absolute space is the reference frame of this aether. But Einstein's theory does not incorporate this model, and in fact is able to dispense with the idea of an aether altogether. We will come back to this example, in Chapter 17, in our discussion of the Michelson–Morely experiment – whose results were substantiated by Einstein's interpretation of the Lorentz transformations, rather than Lorentz' own interpretation[4].

NOTES

1 G. Berkeley, Principles of human knowledge, in: D. M. Armstrong, editor, *Berkeley's Philosophical Writings* (Collier, 1965), p. 42.
2 I have discussed this role in: M. Sachs, Space, time and elementary interactions in relativity, *Physics Today* **22**, 51 (1969).
3 Newton's view of absolute space is discussed in: M. Jammer, *Concepts of Space* (Harvard, 1954).
4 H. A. Lorentz, Michelson's interference experiment, in A. Einstein, H. A. Lorentz, H. Weyl and H. Minkowski, *ibid.*, p. 3.

11

On the Fitzgerald–Lorentz contraction

The assumption made in Chapter 10, in coming to a conclusion about whether or not a swiftly moving stick would fall into a hole in the ice, was that the sole criterion has to do with the (gravitational) interaction relation between the material stick and the Earth, at the location of the hole. [It was tacitly assumed that there would not be any tipping as the stick approaches the hole, so that all of the material components of the stick would fall together or not at all.] This criterion, then, was independent of the states of motion of any observers who might be analysing the stick's motion, from other relatively moving frames of reference. With this criterion, the falling of the stick into the hole is dependent only on its interaction with the Earth below it. Any resistance to the stick's fall below the ice is due to the fact that in the frame of reference of the stick and the hole, together, the extension of the stick is greater than the diameter of the hole – even though it did not seem that way at first glance, to the stationary observer, say a fish looking out of the hole at the approaching stick. Had this fish known about the theory of relativity, it should have been surprised to see that the stick did not fall into the hole. But with sufficient intelligence, the fish may have learned to formulate an abstract theory that might explain its previous conclusion about the shortened length of the stick – it may have led the fish to an independent discovery of the theory of relativity!

A natural question that arises at this stage of the discussion is the following: why should an observer of a moving stick deduce its length to be shorter than its real, physical length, by the factor $[1 - (v/c)^2]^{\frac{1}{2}}$? The key to this answer is in the appearance of the universal speed, c, in this factor.

As we have discussed previously, this universal speed is the speed of propagation of an interaction between matter. When one concludes the locations of the ends of a moving stick, the 'interaction' would be that of the stick and the human being's eyes (or any other instruments that may be used for the measurement). Because it takes a finite amount of time for the signal (the propagating interaction) to travel from the moving object to the point of observation, when the time has been reached for the observer to 'see' the moving object, it is no longer at the location where the signal was emitted. That is, the measurement of the location of an object only gives an apparent location. To determine the actual location of the object at the time when it emitted the signal, at c cm/sec, (the object itself moving at v cm/sec relative to the reference frame of the observer), the observer must apply the Lorentz transformation factor in the correct way – that is, in the way that would preserve the forms of the laws of nature in the relatively moving frames. According to the information gathered from the observer's apparatus, and the deductions she concludes from these measurements about the length of the moving stick, she is given the illusion that the stick is shorter than it really is. However, knowing about the theory of relativity, the observer would then apply the Lorentz transformation formula, thereby projecting herself into the frame of reference of the moving stick to deduce its real length – i.e. the magnitude of its length that is independent of the state of motion of any observer who happens to be looking at it.

This is analogous to our previous deductions about the night sky, though the corrections depending on the speed of light are not the same. We see many stars in distant galaxies with our telescopes, and we know that these stars emitted the light that we see now many years ago. That is, at the time in the astronomer's frame of reference when the star is observed, it is no longer at the location where it was when the signal was emitted. In fact, the star may have burned itself out millennia ago!

The conclusion about the finite time of propagation of interactions does not occur in Newton's theory; his interactions do not take any time to propagate – they occur spontaneously, simply by virtue of the existence of the interacting components at particular locations in space. That is, if the Sun's centre of mass should be shifted by 1000 miles relative to the Earth's location, Newton would have claimed that the Earth should respond to this change immediately – this is equivalent to saying that the interaction travels at an infinite speed. This concept in Newton's theory is called 'action-at-at-distance', and is a fundamental concept in the classical view.

In contrast, the idea of a finite propagation time for the interactions between matter is fundamental in the continuous field approach, as it was originally proposed in the electromagnetic theory and later found to be a necessary ingredient in the expression of the theory of relativity, as a general theory of matter.

Summing up, then, the reason for the appearance of the speed of light, c, in the shortening factor in the Lorentz transformation, giving rise to the 'Fitzgerald–Lorentz contraction', is that this is the speed of propagation of the interaction between material components in a measurement – the observer and the observed. The contraction would disappear if we should assume, with Newton, that this speed of propagation of the interaction is infinite, so that we have 'action-at-a-distance'. But the experimental evidence at this stage of physics reveals that the claim of relativity theory, which superceded Newton's classical view, is the correct approach.

Question The problem of distinguishing between what is real and what is illusory, depending on one's frame of reference, seems to me to be crucial. How might you compare this in physics with the actions of human beings due to prejudices when it comes to our making judgements?

Reply We tend to think that whatever is out there has nothing to do with our own selves, except for the effects on us, as an external action. Thus, if an astronomer might see a happening in the night sky, and judge it to mean that a particular sort of occurrence is taking place, he is aware of the fact that his judgement is based on a particular point of view. That is to say, the judgement should not be taken, at first glance, as an objective statement, until all possible interpretations have been considered – that is, the possible interpretations within the scientists' framework of thinking.

For example, many human beings take it for granted that their own particular classification, within the human race, is inherently superior to other classifications of human beings, in regard to some human quality, such as intelligence. The reasons used by these people in coming to their conclusion may seem perfectly logical, and even quantitatively correct – as far as they go! But such claims tend to use only partial information, that is taken from some 'absolute' frame of reference. According to the interpretation of a relativistic language – to express the different aspects of a single, objective world – the individual who is claiming objective truths must subject them to transformations to all other frames of reference, looking for the features in their relations that remain unchanged, i.e., the 'invariants'. Only then can anyone express any views of the world from an objective point of view – that is, until new information might lead to an abandonment of its objective validity.

With such sensitivity to the objectivity and oneness of the real world, I think the egocentric dominated prejudices of one person or another (or one group of people within the society or another) toward fellow members of the human race would diminish, leading to an improvement of human rights.

In the preceding example, in which a fish in part of a frozen pond looks out of a hole in the ice to see a stick with a bug sitting on it, sliding at high speed towards it, the fish and the bug (initially) came to opposite conclusions about the outcome of a well-defined experimental situation. If they could have communicated with each other, they might have come to the realization that the other was a trustworthy observer and thinker. Realizing, then, that each was coming to an opposite conclusion about the outcome that indeed must be unique (the stick either falls into the hole or it does not!) they would have been forced to re-examine the situation together, until they could arrive at a theory that would satisfactorily explain why it was that each of them came to opposite conclusions, initially, as well as both of them agreeing on a single outcome for the situation. Thus, the bug and the fish might then agree on a single theory, with a unique set of predictions. Such a theory would necessarily be 'abstract' in the sense that it must underlie the observations which, on the surface appear to lead to contradictory conclusions. In this case, it would have been the mutual respect of both observers for the other's point of view, and their mutual faith in the uniqueness of the laws of nature, that would have led to the removal of their respective prejudices, formed from their initial experiences, that had previously prevented them from making progress in their understanding of the real world.

12

Relative time and the twin paradox

Similar to our conclusion about the meaning of the relativity of spatial measures, a correction factor that occurs in the comparisons of time measures in relatively moving frames of reference must also be interpreted, within the theory of relativity, as not more than a language translation – a change of scales in order to maintain the forms of the laws of nature in the relatively moving frames of reference, from any observer's view.

Placing a clock at the spatial origin of a coordinate frame (i.e. where the spatial location is at $x = 0$) the Lorentz transformation formulae imply that the change of scale of time measure must be determined from the equation:

$$t = t' \left[1 - \left(\frac{v}{c} \right)^2 \right]^{\frac{1}{2}}$$

where, as before, v is the relative constant velocity of the reference frames in which the laws of nature are to be compared, according to special relativity theory. Thus, a time interval t, as deduced from a measurement in Bill's frame of reference by Sally (the stationary observer), is shorter by the Lorentzian factor, $[1 - (v/c)^2]^{\frac{1}{2}}$, compared with the corresponding time interval, t', that Sally would compare with a measure on a watch on her own wrist.

This brings us to the famous 'clock paradox', also known as the 'twin paradox'. If it should be true that there is a one-to-one correspondence between the time interval measured on the face of a clock that is in motion relative to a stationary clock, as deduced, say, by Sally, from her measurements, and a physically evolving process in the moving frame of reference, such as the unwinding of the spring of a clock, then Sally would have to claim that the traveller, Bill, as well as the clock in his frame of reference, and all other physically evolving processes in his frame, are ageing at a slower rate than the ageing processes in her own frame of reference. If, for example, instead of Sally and Bill, Paul and Peter are twin brothers, then if Peter should take a trip into outer space, and manage to travel at speeds that are close to the speed of light, c, then when he returns home, one might conclude that he should be noticeably younger than his twin brother, Paul, who stayed at home. His fraction of Paul's age being given by the Lorentz factor, $[1 - (v/c)^2]^{\frac{1}{2}}$. If the ratio v/c is very close to unity, Peter may have not aged at all during the journey, i.e. the time passage t is close to zero, while the time passed for his brother, t' could be many years of ageing! In fact, Peter may return home after his rapid journey to meet his great-great grandchildren, who would be physically older than he is!

The trouble with this conclusion is that, according to the principle of relativity itself, motion is a subjective aspect in the description of the physical laws (as it was originally discovered by Galileo, 300 years before Einstein!). Thus, the twin brother, Peter, in the rocket ship, could just as well have been the one who is called 'stationary', and making measurements in his brother Paul's reference frame, on Earth. That is, he would see his brother Paul travel away from him at $-v$ cm/sec (i.e. moving in the opposite direction from that seen by Paul when he was looking at him in the rocket ship) – Paul later to return to him, at the end of the journey. The same analysis that Paul carried out before could now be carried out by Peter, leading him to the conclusion that after the journey is over, Paul would be much younger than he is – opposite from Paul's conclusion, but by the same time contraction factor, $[1 - (v/c)^2]^{\frac{1}{2}}$.

Had Peter relayed all of his data to Paul during the journey, then with Paul's confidence in Peter as a competent scientist, Paul would have had to agree with Peter's conclusion. Then, Paul would have to conclude that at the end of the round-trip journey his brother Peter would be both older and younger than himself! This is a logical paradox – it is called the 'twin paradox'.

The same analysis applied to physically identical clocks, initially synchronized, then implies that after the round-trip journey of one of them from the other, the clocks would lose their synchronization, but one of them would be both slow and fast compared with the other! – this is the 'clock paradox'.

The appearance of a logical paradox in the outcome of the predictions of a theory means that there was a false or inconsistent interpretation of the theory, somewhere along the way. The resolution of the paradox may come from the observation that the time reduction factor, $[1 - (v/c)^2]^{\frac{1}{2}}$, came from the comparison of time measures of reference frames that are in uniform motion with respect to each other (i.e. constant relative velocity, in a straight line), while for the round-trip journey one twin must accelerate away from his brother, before settling down to the constant speed v, and then decelerate in turning around for the return journey, then decelerating once again in coming to rest with respect to his brother, at the end of the journey. Thus, one twin brother makes non-uniform changes, three times during the round-trip journey. During these periods of non-uniform motion, the Lorentz transformations are not valid. Perhaps, then, this implies a non-equivalence in the relative motions of Paul and Peter that would yield the same conclusion for both of them – either one of them is younger than the other after the journey is over or there is no asymmetric ageing effect at all.

But even when we take account of the periods of non-uniform motion, there is still trouble from the point of view of relativity theory. For just as Paul would claim that Peter is accelerating relative to him, so Peter would claim, from his perspective, that Paul is accelerating away from him (in the opposite direction). If one should assume that acceleration is an absolute effect, independent of time, as caused by an external force, then one would have to conclude that the force that caused the rocket ship to leave the Earth, turn around, and then stop at 'splash down', caused absolute acceleration effects for Peter and not for Paul, thus introducing an asymmetry in the motion – possibly being responsible for the asymmetric ageing effect that is claimed! On the other hand, according to the assertion of the principle of relativity, space and time are purely relative concepts, so that motion of any type (uniform or non-uniform) is a purely subjective entity. Since motion of any type is relative, both Peter and Paul must have descriptions of their observations of each other, from their respective frames, that are in accordance with the corresponding laws of nature – i.e. each should claim the same sort of observation when looking at the other. Thus, the paradox remains if we interpret the change of time scales to correlate directly with ageing.

The fact that acceleration entails force (as a cause) does not imply, as many claim, that there is an absolute reference frame associated with non-uniform motion. For any action of Paul on Peter, there is a reaction of Peter on Paul. Further, the previous conclusion about asymmetric ageing only entailed the use of the Lorentz transformation formula – no explicit force, that might predict the asymmetric ageing effect, came into the description! It is rather a prediction of a physical effect, without a physical cause. I must then conclude that the claim of asymmetric ageing is logically invalid with respect to the theory of relativity itself, which starts with the assertion about the objectivity of the laws of nature – the physical cause–effect relations[1].

In the early stage of his career, Einstein himself believed that the traveller, Peter, would return to Earth younger than his twin brother. His reason was based on the observation that Peter and Paul are not actually in symmetrical frames of force. This is because, from the view of Paul, Peter leaves him and a stationary planet Earth, along with a particular spatial distribution of stars, planets, and the rest of the entire universe. On the other hand, from Peter's view, Paul leaves him (in the opposite direction) along with Earth, (as well as the rest of the universe), i.e. there are not only Paul and Peter to consider, there is also the Earth's motion to consider, from Peter's view, and not from Paul's. Thus Peter's and Paul's motions are not symmetrical; their physical conclusions must then necessarily differ in regard to what they will deduce from their observations. Einstein then showed that, with this inequivalence of the motions of Peter and Paul, and arguments based on the 'equivalence principle' of general relativity theory, (which is the assertion of an equivalence of gravitational forces on a stationary body and a uniformly accelerating body in free fall), that when account is taken of the way the total motions are described, both Peter and Paul would then agree that it would be Peter (not Paul) who ages less during the round-trip journey.

There are logical objections to this conclusion – that it is contrary to the meaning of space and time in relativity theory, as not more than subjective language elements that are there only to express the laws of physical processes, rather than being the physical processes themselves! For the asymmetric ageing effect we have discussed is indeed a physical process itself. The questions then arises as to whether or not relativity theory truly implies the existence of a one-to-one correspondence between a fixed observer's time measure for a moving clock and the physical ageing of that clock, such as the actual unwinding of the spring of a clock (or the irreversible cell decay associated with the ageing of a human being).

There is also a further objection to Einstein's resolution of the twin paradox, whereby he uses the equivalence principle of general relativity. This is that the theory of relativity should apply to all possible physical situations – whether or not they presently exist. Consider, for example, a fictitious universe composed soley of the twins Peter and Paul, each in a separate rocket ship. Suppose that initially they move together with no relative speed between them and no mutual potential energy of interaction. That is to say, they are initially in the same inertial frame of reference. Suppose now that at some time (in their common coordinate frame) Paul should blast off his rocket motors, parting company with Peter's rocket ship, but returning again to Peter's ship at some later time. In this universe, there would be no large planet, Earth, to appeal to, to resolve the paradox. The motions of Paul and Peter, including their relative accelerations, would be totally symmetrical and the paradox would remain.

In addition to philosophical objections to the 'reality' of the twin paradox, it might be remarked that in spite of voluminous writings on this particular problem (since 1905!) there has been no conclusive evidence to support the alleged asymmetrical ageing effect. There have been numerous experimental verifications of the prediction that the time measures of relatively moving clocks must be calibrated according to different scales, from the view of any particular observer of physical phenomena in the respective frames of reference, in accordance with the Lorentz transformation formula. But there is no direct experimental confirmation to support the alleged effect of asymmetrical ageing, such as an actual difference in the unwinding of the springs of compared clocks in relatively moving reference frames, or the compared decays of unstable particles, leading to different numbers of particles in the reference frames, compared with the initial state whereby the samples in the same inertial reference frame contained the same number of particles. For example, consider two boxes, each containing N radioactive nuclei, and one of them is taken on a round-trip journey at high speed. When it returns to the site of the box that did not take the trip, would there be more remaining nuclei in the travelling box than the stationary box? – implying that the travelling box becomes younger than the stationary one. There is no experimental evidence for such an effect.

But even if there would be such evidence, this is beside the point made here – that the theory of relativity, *per se*, does not predict such an effect, one way or the other. It does say that if there should indeed be forces in nature that cause the asymmetrical ageing effect, due to the relative motion, then the laws of these effects must conform with the principle of relativity. This is all that this theory says about it – it does not imply that such an effect, and the law that accompanies it, necessarily exists.

Question Aside from the lack of conclusive experimental data to claim the asymmetric ageing effect, you said that there is no theoretical prediction from the theory of relativity that the effect really occurs. Is there any theoretical proof that the asymmetric ageing effect is not predicted by the theory of relativity, *per se*?

Reply I have completed, and published, a theoretical proof that I claim shows, unambiguously, within the full mathematical structure of the theory of general relativity, that asymmetric ageing is not predicted by this theory, *per se*[1]. The details of this proof are beyond the level of this book. Still, I might mention what assumptions went into this proof.

Firstly, in agreement with Einstein, if the effect exists at all, it must be a consequence of general relativity – which contains the mathematical description of special relativity as a limiting case. The principle of general relativity implies that frames of reference are distinguishable only in terms of relative motion of the most general type, as we discussed previously. 'Motion', in turn, is defined only in terms of continuous transformations in space–time. I have found in my research programme that while Einstein's original tensor equations satisfy the principle of relativity regarding the continuous transformations, they also maintain their form with respect to the (discontinuous) reflections in space–time. This is an extra symmetry that is not required by the principle of relativity – which underlies the theory in the first place. The removal of the discrete reflection symmetry elements, while maintaining only the continuous transformations, then led to a more general mathematical expression of general relativity theory. This more general form then provided a mathematical expression that permits an unambiguous resolution of the twin problem, but without the need to construct special models, or the need to use approximations as we have seen in a great deal of the literature since special relativity was first presented to the scientific public.

In carrying out the proof, one starts out by correlating the 'proper time measure', over a particular path in space–time – a measure that is calibrated as a standard, in accordance with some physical ageing process that is evolving in this space–time reference frame, i.e. one sets up the 'proper time' as a scale of measure within the reference frame that describes the physically ageing system, such as the cellular degeneration of a human body or the unwinding of a steel spring of a mechanical clock, or the decay of unstable elementary particles, or whatever physically evolving process one wishes to use for a calibration of the time measure, as a standard. We then consider two physically identical clocks (of any type). If these two clocks should have different histories in space–time, while starting out relatively at rest in the same inertial frame of reference, and rejoining later on at a different space–time point, once again at rest in the same inertial frame, and if the intervening different paths in space–time are equal in length – so that the total proper time passed by the clocks on the respective paths are the same – then the correspondence of the proper times with the ageing of the respective clocks implies that there should be no difference in the ageing of these physical systems. In this way, it was proven that general relativity theory, when fully expressed, implies that there should be no asymmetric ageing simply by virtue of the relative motions of the respective physical systems.

Of course, this result assumes that no external forces came into the picture, such as a very forceful blast-off of a rocket ship, thereby destroying the synchronization with another physically identical clock on the ground at the launching site. But the proponents of the asymmetric ageing effect do not propose any extra forces (i.e. cause–effect relations) to account for the effect. The claim is that asymmetric ageing occurs only because of the difference in time scales of the relatively moving frames of reference! It is this conclusion that I claim to have disproven in an unambiguous, mathematical fashion, based on the most general type of formulation of this theory, consistent with its underlying principle of general relativity.

Question Would the proponents of the idea that relativity theory predicts that the twins should age asymmetrically also claim that they would be different sizes at 'splash down'? – because of the Fitzgerald–Lorentz contraction?

Reply If the logic about asymmetric ageing would be correct, then due to the fact that the traveller is not really a point at the origin of the coordinate system, he would also be expected to reduce in size, in the direction of his motion – just before he stops his journey! Thus if the travelling twin should move fast enough, he may return to his brother not only thirty years younger, but just before stopping the journey he may be only two inches tall, while his brother is still six feet tall! – assuming that he was oriented in the direction of his travel. If his position in the rocket ship had been perpendicular to his motion, then when he returns to his twin, just before splash-down, he may still be six feet tall but only two inches thick – a wafer! In fact, had he travelled much closer to the speed of light than this, those who adhere to this view might claim that the traveller would return just a few microns thick – his brother would not be able to see him if he should stand sideways – just before their reunion. But when the traveller would come to an absolute halt, he should suddenly restore his original size, according to this view.

I do not believe that these physical effects are in fact predicted by the theory of relativity for the same reason that I do not believe that the theory predicts asymmetric ageing. There is no cause–effect relation within the theory that demonstrates an explicit force (physical cause) – such as the exertion of a pressure on the traveller's head and feet that would cause his body to shrink to two inches tall! To draw the conclusion of a physical effect without the existence of a physical cause is illogical – according to the scientific basis of the theory of relativity, itself!

Question You mentioned that, in his earlier period, Einstein did believe that the travelling twin would come home physically younger than his brother, by the factor that appears in the Lorentz transformation formula. In view of the logical error that you pointed out, did Einstein change his mind about this in his later years, i.e. after his 1905 discovery of special relativity?

Reply I haven't seen any direct recanting of his original opinion on this. But from what he did write after that time, it seems to me that indeed he did change his mind about it, as his theory matured. I have a few quotations, that you might consider, to back my opinion. In a lecture that Einstein gave to the Prussian Academy of Sciences, in 1921, he said (my italics)[2]:

> Geometry predicates nothing about relations of real things, but only geometry together with the purport of physical laws can do so. . . . The idea of the measuring rod and the idea of a clock coordinated with it in the theory of relativity *do not find their exact correspondence in the real world*. It is also clear that the solid body and the clock do not in the conceptual edifice of physics play the part of irreducible elements, but that of composite structures, *which may not play any independent part in theoretical physics*.

In his *Autobiographical Notes*, that he wrote in 1947, 26 years later, Einstein said[3]:

> . . . strictly speaking, measuring rods and clocks would have to be represented as solutions of basic equations (objects consisting of moving atomic configurations), not, as it were, as theoretically self-sufficient entities.

Taking the latter statement of Einstein at face value, it should then follow that to predict the actual reading of a clock – the locations of its hands – or the length of a material rod – the extension of its constituent molecules, relative to each other – in a reference frame that is in motion relative to an observer, compared with the reading of the clock and the length of the rod in the observer's own frame of reference, one must necessarily appeal to laws of matter (i.e. the dynamical cause–effect relations) that predict the behaviour of moving material systems of matter, called 'rod' and 'clock'.

Einstein then went on to say[3], (my italics):

> if one did wish to forego a physical interpretation of the coordinates in general . . . it was better to permit such *inconsistency – with the obligation, however, of eliminating it at a later stage.*

Thus we see that Einstein admitted that a logical inconsistency is indeed encountered if one should interpret the kinematic space–time transformations of the theory of relativity operationally, in terms of dynamical cause–effect relations, as he did himself in the early stages of development of the theory. Still, he said that if one should use this inconsistent interpretation, as a temporary tool, he had the obligation of eliminating it at a later stage of the theory.

The latter assertion indicates to me, in no uncertain terms, that, after his initial claims about physical implications of the Lorentz transformations – such as the claim that a twin brother should return from a high speed, round-trip journey, physically younger than his twin brother – he did indeed change his mind about this conclusion, no longer taking seriously the earlier interpretations of the Lorentz transformations as having absolute, dynamical consequences. He rather saw them as strictly descriptive, kinematic relations, prescribing the rules for translating the language parameters from one reference frame to another in order to preserve the objectivity (covariance) of the expressions of the laws of nature – thereby eliminating the problem of logical inconsistency in the theory of relativity.

NOTES

1 I have reviewed the logical argumentation about the twin paradox in: M. Sachs, "A resolution of the clock paradox", *Physics Today* **24**, 23 (1971). In this article, a mathematical argument is also presented that refutes the claim of asymmetric aging due to relative motion.

2 A. Einstein, *Sidelights of Relativity* (Dover, 1983), p. 35.

3 P. A. Schilpp, (editor) *Albert Einstein – Philosopher–Scientist* (Open Court, 1949).

13

Geometry, causality and the light cone in special relativity

Geometry is a system of logical relations imposed on families of lines and points. With this set of relations as 'freely invented' axioms, one can logically prove that further relations must follow. Euclid's geometry is a set of logical relations that were hinted by the human being's perceptions of a 'three-dimensional space'. But geometry, *per se*, is not a law of nature. Rather, it has been useful in describing the properties of matter, since the times of ancient Greece, so useful that natural philosophers thought that Euclid's geometry must necessarily underlie the 'science of space' of the physical universe.

Nevertheless, there is nothing to stop one from inventing a different set of axioms, not all of them being compatible with Euclidean geometry. In the 19th century, a few non-Euclidean geometries were invented. One of these, 'riemannian geometry', was found to be useful by Einstein in his generalization from special relativity (to be discussed in Chapters 18 and 19), that, in turn, led to our present understanding of the gravitational force.

The following 'axiom of parallels' is an example of an axiom from Euclidean geometry:

> Given a straight line *L* and a point *P* outside of it, one and only one other straight line, *L'*, in the plane of *L* and *P*, going through *P*, will never intersect *L*.
> The two lines, *L'*, *L* are then said to be 'parallel'.

In the statement of this axiom, 'straight line' is defined as the locus of points such that the distance between any two of them must be a minimum. Such a locus is generally called a 'geodesic'. Neither riemannian geometry nor the other non-Euclidean geometries discovered in the 19th century contain this 'axiom of parallel lines'.

The four-dimensional 'metric' of space–time in Einstein's theory of special relativity obeys the axioms of Euclid's geometry. A 'metric relation' is a particular restriction on the points in a (general) space, e.g., in ordinary 3-dimensional space, all points on the surface of a sphere project into one another when the sphere is rotated by any angle about any axis through its centre. Alternatively, we can say if (x, y, z) are the coordinates that locate a point on the surface of a sphere, then when the coordinate frame is rotated by an arbitrary amount, these spatial coordinates rotate into other spatial coordinates on the surface of the sphere as follows: $x \rightarrow x'$, $y \rightarrow y'$, $z \rightarrow z'$, such that

$$x^2 + y^2 + z^2 = x'^2 + y'^2 + z'^2 = \text{constant}$$

The constant on the right-hand side of this equation is the square of the radius of the sphere. This equation for the radius of the sphere then represents a restriction on the coordinates that is an example of a 'metric relation' in the 3-dimensional space.

There is a similar sort of metric relation in the space–time of special relativity theory, except that it involves four coordinates (x, y, z, ct), rather than three. As an example, suppose that a light signal should be found by some observer, in his own frame of reference, to be emitted at the place (x_1, y_1, z_1) at the time t_1, and to be absorbed at the place (x_2, y_2, z_2) at the later time t_2. Since the electromagnetic interaction (light) travels at the speed c in a vacuum, it must travel the distance $c(t_2 - t_1)$ between the emitter and absorber. According to the measurement from this frame of reference, then, the temporal measure of distance, $c(t_2 - t_1)$, must be equal to the spatial distance travelled, i.e.

$$c(t_2 - t_1) = [(x_2 - x_1)^2 + (y_2 - y_1)^2 + (z_2 - z_1)^2]^{\frac{1}{2}} \tag{13.1}$$

This equation may then be re-expressed in the form

$$(\delta s)^2 = c^2(\delta t)^2 - (\delta r)^2 = 0 \tag{13.2}$$

where (δr) is the spatial interval on the right-hand side of (13.1) and (δt) is the corresponding time interval $(t_2 - t_1)$.

Suppose now that in some other frame of reference an observer $0'$ records the same emission and absorption of light. In her language the light signal leaves the place (x_1', y_1', z_1') (at the emitter) at the time t_1', and arrives at the absorber at the place (x_2', y_2', z_2'), at the later time t_2'. With the assumption that the speed of light is independent of the frame of reference in which it is represented, it follows that $0'$ would claim a version of (13.2) expressed as follows.

$$(\delta s')^2 = c^2(\delta t')^2 - (\delta r')^2 = 0 \tag{13.3}$$

Noting that the right-hand sides of (13.2) and (13.3) are both equal to zero, we have the equality of four-dimensional intervals:

$$\delta s = \delta s' \tag{13.4}$$

With respect to the path of the light signal, then, the magnitude of the four-dimensional interval (δs) is independent of the space–time frame of reference in which it is expressed. The suggestion then appears that perhaps for any sort of motion, not necessarily that of a light signal, it may still be true that these intervals are equal in all inertial reference frames.

Following this suggestion, we see that this is indeed the case when we insert, for example, the following Lorentz transformations for the 'primed' variables:

$$\delta x' = \frac{\delta x + v\delta t}{\left[1 - \left(\frac{v}{c}\right)^2\right]^{\frac{1}{2}}}$$

$$\delta y' = \delta y \qquad \delta z' = \delta z$$

$$\delta t' = \frac{\delta t + v\dfrac{\delta x}{c^2}}{\left[1 - \left(\frac{v}{c}\right)^2\right]^{\frac{1}{2}}}$$

into the interval

$$(\delta s') = c^2(\delta t')^2 - (\delta x')^2 - (\delta y')^2 - (\delta z')^2$$

For it then follows that

$$(\delta s')^2 = c^2(\delta t)^2 - (\delta x)^2 - (\delta y)^2 - (\delta z)^2 = (\delta s)^2$$

An equivalent way of saying that the laws of nature preserve their forms under the transformations between relatively moving inertial frames of reference is then to assert that such coordinate changes must be those that leave the four-dimensional interval δs unchanged.

Analogous to our previous definition of the metric of a three-dimensional space in terms of the projection of the points of a sphere onto each other, under the rotations of the sphere with constant radius, special relativity entails instead the four-dimensional space–time containing the 'metric' restriction that the 'four-dimensional radius' $\delta s = [c^2(\delta t)^2 - (\delta r)^2]^{\frac{1}{2}}$ must remain constant. The constancy of this four-dimensional interval corresponds, geometrically, to the equation for a cone, and is called the 'light cone'.

CAUSALITY

Causality may be defined within the geometrical approach in the following way. Suppose that it should be possible to find a frame of reference of space and time coordinates in which two events happen at the same place, but at different times. Calling this the 'primed' frame, $(\delta r')^2 = 0$, $\delta t' \neq 0$. In this case,

$$(\delta s')^2 = c^2(\delta t')^2 = c^2(\delta t)^2 - (\delta r)^2$$

where the 'unprimed' coordinates refer to any other inertial frame of reference. Since the left-hand side of this equation (being a square) is a positive number, the right-hand side must also be positive. Thus, the time interval $c(\delta t)$ must be greater than the spatial interval (δr).

Such intervals of space–time are called 'timelike'. They describe a signal propagating between two spatial points whose separation is less than $c(\delta t)$. One may then look at this emission and absorption of a signal in terms of cause (emission) and effect (absorption). The events at the two spatial endpoints separated by (δr) are then said to be causally connected.

Suppose, on the other hand, that it should be possible to find a coordinate frame in which two events take place at the same time, but at different places. In this case, $(\delta t')^2 = 0$, $(\delta r') \neq 0$ Thus, for this situation,

$$(\delta s')^2 = -(\delta r')^2 = c^2(\delta t)^2 - (\delta r)^2$$

Since $-(\delta r')^2$ is a negative quantity, the right-hand side must also be so. In this case, the spatial separation (δr) is greater than $c(\delta t)$. Thus the signal emitted at the spatial point r_1 and travelling at the maximum speed, c, does not have enough time in the interval δt to reach the spatial point r_2. Such four-dimensional intervals are called 'spacelike'. They cannot be identified with cause–effect relations because the spatial separation is too great for the signal to go from one end (cause) to the other (effect) in the time (δt), at the speed c.

The 'timelike' intervals are inside or on the light cone. The 'spacelike' intervals are outside the light cone. Thus, the light cone is a geometrical configuration of causally connected points in space and time, between the present (the apex of the cone) and the future (the points in the cone with positive t) and the past (the points in the cone with negative t). These are the 'forward' and 'backward' light cones, respectively.

Question In classical physics, causality usually refers to the earlier cause and the later effect, that is, it refers to a temporal ordering of events. But, according to what you have said, time no longer has this same meaning in relativity physics. How does this change in the meaning of time affect the concept of causality within the logical basis of relativity theory?

Reply As you comment, one should be cautious in extending the concept of causality from classical physics to relativity physics, because of the revolutionary change that occurred in regard to our interpretation of time. According to our previous discussions, space–time in relativity theory is not more than a coordinate frame that serves the purpose of facilitating an expression of the laws of nature. The laws of nature, in turn, refer to an ordered set of relations that are supposed to underlie the way in which we must comprehend the universe. It is the latter set of relations that are supposed to be the cause–effect connections, according to this theory.

Clearly, this definition does not single out the time coordinate in particular, as the classical theory does. In classical physics, time is an absolute, objective entity, and allows causality to be defined in terms of an ordering of 'absolute events' in time.

What is elementary in relativity theory is the objectivity of the relations that underlie the physical features of a closed system. We have seen that a useful language to describe causal relations in relativity theory entails a blocking out of all intervals in space–time that are 'spacelike' – that is, intervals outside the light cone. In contrast, classical physics uses the entire space and time – essentially because of the underlying concept of 'action-at-a-distance' and the accompanying absoluteness of the time measure.

Because of the more general definition of cause–effect in relativity theory, as related to a set of logically connected features of a closed system – not necessarily in a temporal order – causality, in this theory, is synonymous with 'determinism', in the sense of asserting the existence of a predetermined order for any physical system. The cause–effect concept of classical physics is only a special case of the more general concept in relativity theory – it corresponds to taking a special frame of reference in which a particular observer's time coordinate is singled out in his expression of the physical laws, in his particular coordinate frame of space and time parameters. With this more abstract view of causality, we also see a different way of interpreting the 'backward' light cone – as an abstract portion of the language used to represent 'retrodiction' in the physical laws – indicating events in the past that are causally connected to the present. It is nothing more than this. [This meaning negates the erroneous claims of some that relativity implies that we may physically move into the past!] This is symmetric with the 'forward' light cone, that is similarly a picture of the causal connections of the present with the future in the form of 'prediction'.

Question How would the causality concept in relativity apply to the description of the human society, rather than a society of inanimate things, as in physics?

Reply One might compare the feature of the mathematical relations having to do with causal connections within a physical system, with the pertinent range of communications (signals!) between members of the human society. If an individual person, Tom, should fail to communicate with another individual person, Akira, the reason may be the outward differences in their respective ways of thinking. Because of this lack of communication, Tom (or Akira) may then conclude that Akira (or Tom) is evil, inferior, and so on. But such feelings are irrational – they are not causally connected, following rationally from initial premises. When such conclusions and others, more subtle perhaps but equally irrational, are recognized as irrational, then they should be rejected as not relating to reality, which is the actual mutuality of the Tom–Akira relation. This is analogous to the idea that the space–time intervals outside of the light cone are blocked out as not causally connected.

To make the attempt to understand one's neighbour, irrespective of outward differences, the individual must successively build up toward closer realization of mutually overlapping concepts. The interaction between persons then becomes a rationally connected relation, analogous to the causally connected points of the light cone. To proceed in the direction of attempting to understand one's neighbour, the human ego must accept the forsaking of the concept of a human being as a 'thing-in-itself', replacing it with an entity involved as an inseparable component of a closed system. This may be a system that outwardly appears as two or ten or three milion components of the society – but a society that is in fact without actual separable parts. It is a 'holistic' society.

The example of 'two' is perhaps the simplest to study. In the human society, a poignant manifestation of 'two' is the love that is shared by a man and a woman. When it is a sincere love, the participants experience, at least in part, a relinquishing of their apparent individualities to a single mutuality – an entity that one might call, man–woman.

It appears to me that the oneness of the human love relation is conceptually comparable with the oneness of man, the universe and God, according to Martin Buber's *I–Thou* philosophy. This is a holistic relation that is without parts. It is not set in the context of space and time. It is 'pure relation' without the separate relata. It is expressed in the form of a hyphenated word, – in which the hyphen is, by definition, irremovable. Buber's philosophy also incorporates the manifestation of the closed system in which the human consciousness can place itself in a state that is very weakly coupled to the remainder of the closed system, seeming to be approximately disconnected from it. In this form, which Buber calls 'I–It', the relation does seem to be in terms of separable parts, and set in the context of space and time. Indeed, when a true love between a man and a woman, an *I–Thou* relation, should degenerate to a loss of the full oneness, then each of the components (the man and the woman) become independent constituents. They start to see themselves as *I* (the subject of the relation) and *It* (the object of the relation). This is without the symmetry in which the interchanging of the object and the subject preserves the love relation (as in a covariant relation according to Einstein's principle of relativity). One must recover the symmetry between subject, *I* and object, *Thou* in the *I-Thou* relation, as a single entity – similar to the closed system model that is implied in physics by Einstein's theory of relativity. At least, this is how I see some of the connections in relativistic physics with the problem of society.

14

Particles of matter in special relativity and $E = Mc^2$

When Einstein applied the transformations of space and time coordinates that would preserve the laws of nature in relatively moving inertial frames (i.e. where the relative motion corresponds to constant speed in a straight line) to the laws that govern the motions of particles of matter, called 'relativistic mechanics', new predictions followed that were not implied by Galileo's relativity. One of the most famous of these new predictions, that differed with the classical theory, was that when a body with mass M is at rest and not in any field of potential energy, it still has a residual energy – the 'rest energy', $E = Mc^2$. The original derivation of this relation was quite straightforward, and it demonstrates the intimate relation between the particle aspect of matter and its continuous field aspect. The following is an outline of this derivation[1].

Consider a quantity of radioactive matter, with inertial mass equal to M, spontaneously emitting electromagnetic radiation (gamma rays). Suppose that before this matter emits a quantity of radiation, an observer, Ivan, in the frame of reference of the matter (the 'proper' frame) measures the total energy of this matter to be E_0 units. [Recall that in physics, 'energy' is the quantitative capability of the matter to do work.] After this matter emits E_γ units of radiant energy, as measured by Ivan, he determines that its energy had decreased to E_1 units. According to the law of conservation of energy, Ivan's measurement should then verify that $E_0 = E_1 + E_\gamma$, i.e. the total energy loss of the quantity of matter that radioactively emitted the gamma ray is

$$E_0 - E_1 = E_\gamma \tag{14.1}$$

Suppose now that a different observer, Enrico, travels with the constant, straight line motion, with the speed, v, relative to Ivan's coordinate frame. Enrico would then measure the matter's initial energy to be E_0' and its final energy (after it emits the gamma ray) to be E_1', while the electromagnetic energy radiated away, in Ivan's reference frame, according to Enrico's view, is equal to E_γ'. According to the principle of relativity, then, the energy conservation law in Enrico's reference frame must be in one-to-one correspondence with the energy conservation law in Ivan's reference frame, i.e.

$$E_0' - E_1' = E_\gamma' \tag{14.2}$$

It follows from the comparison of the forms of Maxwell's equations in Enrico's and Ivan's frames of reference, and the expression for the electromagnetic radiant energy (gamma rays) according to the Maxwell formalism, that the transformations between the expressions for this radiant energy in the two inertial frames is as follows.

$$E_\gamma' = \frac{E_\gamma}{\left[1 - \left(\frac{v}{c}\right)^2\right]^{\frac{1}{2}}} \tag{14.3}$$

Combining (14.1), (14.2) and (14.3), it then follows that

$$(E_0' - E_0) - (E_1' - E_1) = (E_\gamma' - E_\gamma) = E_\gamma \left[\frac{1}{\left(1 - \left(\frac{v}{c}\right)^2\right)^{\frac{1}{2}}} - 1\right] \tag{14.4}$$

We first note, in regard to (14.4), that the only difference between E_0' and E_0 is that the quantity of energy, E_0' is measured in a moving platform by the observer, Enrico, while the energy E_0 is measured by the observer, Ivan, who is at rest with respect to the radioactive matter. Thus, the difference $(E_0' - E_0)$ must represent the 'energy of motion'. This is the kinetic energy of matter, as it would be conveyed to Ivan by Enrico, before the emission of the radiation. Similarly, $(E_1' - E_1)$ must represent the kinetic energy of the matter after the radiation had been emitted. The left-hand side of (14.4) then represents the difference in the kinetic energies, as conveyed by Enrico to Ivan (who rides with the matter) from the time before to the time after the gamma radiation was emitted.

In classical physics, and in special relativity theory, when the relative speed of an inertial frame relative to another inertial frame, v, is small compared with the speed of light, c, the kinetic energy associated with the matter in motion is $\frac{1}{2}Mv^2$ in classical physics and approximately so in relativity physics. With this assumption that the relative speed of the reference frames is small compared with the speed of light, the left-hand side of (14.4) then has the approximate form:

$$(\tfrac{1}{2}Mv^2)_0 - (\tfrac{1}{2}Mv^2)_1$$

Since v relates to Ivan's (controllable) motion, relative to Enrico's coordinate frame, it may be arranged that Ivan's speed remains constant from the time before to the time after the gamma ray had been emitted. For example,

Ivan may be moving (with the radioactive sample) in a rocket ship at a constant relative speed, throughout the entire time of these measurements, as controlled by Enrico at the control centre on Earth.

The difference in kinetic energies on the left-hand side of (14.4) is not zero because the right-hand side of this equation is not zero. Since Ivan's speed does not change during the emission of the gamma ray, the difference in kinetic energies

$$(\tfrac{1}{2}Mv^2)_0 - (\tfrac{1}{2}Mv^2)_1 \tag{14.5}$$

which is the right-hand side of (14.4) for v being much less than c, can only be non-zero if there is a change in the mass of the radiating material during its process of emission. Thus, this difference can be expressed as

$$\tfrac{1}{2}(\delta M)v^2$$

where (δM) is the mass change, $M_0 - M_1$.

With the same approximation of small speed of Ivan's frame of reference relative to the observer, Enrico, it follows that one may make the following approximation (using the binomial expansion from algebra).

$$1 \Big/ \left[1 - \left(\frac{v}{c}\right)^2\right]^{\tfrac{1}{2}} \simeq 1 + \tfrac{1}{2}\left(\frac{v}{c}\right)^2 .$$

In this case, the right-hand side of (14.4), along with the approximation of using (14.5) for the left-hand side of (14.4), gives

$$\tfrac{1}{2}(\delta M)v^2 = \tfrac{1}{2}\left(\frac{E_\gamma}{c^2}\right) v^2$$

Solving for the energy of the gamma ray, we have

$$E_\gamma = (\delta M)c^2 \tag{14.6}$$

Thus, according to special relativity theory, when a particle with inertial mass M emits E_γ units of electromagnetic energy, its inertial mass decreases by the amount $\delta M = E_\gamma/c^2$. It then follows that if all of the mass M of this matter should be converted into radiant energy, it would amount to the quantity

$$E = Mc^2$$

That is, Mc^2 is the maximum available intrinsic energy, (i.e. the capacity to do work) of a quantity of matter that has mass equal to M – as would be measured by Ivan, the observer in the rest frame of this matter. For this reason, Mc^2 is called the 'rest energy' of the matter.

However, from the point of view of Enrico, who watches the radiating matter moving relative to him at a constant speed in a straight line, its total energy, according to (14.3), is (in the 'primed' language):

$$E' = \frac{Mc^2}{\left(1 - \left(\frac{v}{c}\right)^2\right)^{\frac{1}{2}}} \tag{14.7}$$

These results are to be compared with the classical result, which implies, instead, that if no external force should be applied to that matter, its energy in the rest frame (measured by Ivan, who moves with it, relative to Enrico) is zero – quite a bit smaller than Mc^2! In the moving frame, according to Enrico's measurement, the classical prediction for Enrico's measurement of the matter's energy would be its kinetic energy, $E' = \frac{1}{2}Mv^2$.

When the ratio v/c is small compared with unity, we have the (binomial) approximation,

$$\frac{1}{\left[1 - \left(\frac{v}{c}\right)^2\right]^{\frac{1}{2}}} \simeq 1 + \frac{1}{2}\left(\frac{v}{c}\right)^2$$

With this, Enrico's measurement of the energy of the moving matter, according to relativity theory, should then be, from (14.7):

$$E' \simeq Mc^2\left[1 + \frac{1}{2}\left(\frac{v}{c}\right)^2\right] = Mc^2 + \frac{1}{2}Mv^2$$

In this case, then, the total energy is the kinetic energy of the matter (as is the case in classical physics) plus the rest energy of the matter, Mc^2. Since the relative speed of the reference frame, v, is much smaller than the speed of light, c, in these considerations, the rest energy term is much greater than the kinetic energy term. [The release of the rest energy of a bread crumb could light up New York City for some time!] Why, then, hasn't anyone ever detected this huge quantity of dormant energy in any seemingly inconsequential bit of matter?

The answer is that one does not directly observe energy, *per se*. Rather, any interaction involved in measurements has to do with energy transfer, which is a difference between two quantities of energy. In the classical experiments, for example, where the kinetic and potential energy convert into one another when a block slides down an inclined plane, the same quantity of rest energy, Mc^2 appears in both of the energy terms that are subtracted from each other, thus cancelling out in the energy difference.

Most people today are aware that nuclear physics experimentation has led to a verification of Einstein's result, with the conversion of the rest energy into other forms of energy. One way to release this energy is to excite a heavy nucleus, such as uranium, by its bombardment with lighter particles, such as neutrons. After absorbing the neutrons, the heavy nucleus becomes unstable, and eventually splits up into several lighter nuclei – called 'nuclear fission'. The sum of the masses of the lighter nuclei that are the fission products is found not to equal the mass of the original heavy nucleus that they came from. The difference, δM, had been converted in this process from the binding energy of the fissioned components into their kinetic energies, equal in magnitude to δMc^2. In this way, a part of the rest energy of the heavy nucleus is utilized when the kinetic energies of the fissioned fragments of the nucleus convert into heat energy.

Another important difference in the description of the mechanical properties of matter, between the classical and the relativity theories, is in their respective predictions for the momentum of moving matter. According to the theory of special relativity, the momentum vector for a quantity of matter with inertial mass equal to M, that is moving at a constant speed in a straight line, with respect to the stationary observer's view, is

$$\mathbf{p}' = \frac{M\mathbf{v}}{\left[1 - \left(\frac{v}{c}\right)^2\right]^{\frac{1}{2}}} \tag{14.8}$$

The classical expression for the momentum of this body is

$$\mathbf{p}' = M\mathbf{v}$$

According to the observer, Ivan, who is at rest with respect to this matter, its momentum is zero, both in classical physics and in relativity physics. However, it is clear that the magnitude of the momentum of matter, to the stationary observer of the moving matter, with v/c near unity, is much greater in relativity physics than it is in classical physics. For with v/c close to unity in (14.8) for the relativistic momentum, the denominator is close to zero, so that the momentum approaches an infinite value.

It is important to note that the reason that the observer of the moving matter concludes that its momentum becomes infinitely large as its speed approaches the speed of light is not that the matter's mass becomes infinitely large! It rather has to do with the way that the stationary observer receives information about the swiftly moving matter relating to its momentum – information that is propagating at the speed of light. Similarly, the expression for energy, from the stationary observer's view of the moving matter, becomes infinite as the speed of the matter approaches the speed of light, when the denominator in the energy expression, given in (14.7), becomes zero. But this is not intrinsically infinite energy! In the frame of reference of the matter itself, (that is, from Ivan's view), the intrinsic energy of the matter is its 'rest energy', Mc^2.

Regarding the particle of light itself, the 'photon' – if this is an adequate description of light – the theory of relativity predicts that it must travel at the universal speed – the speed of light in a vacuum, c – relative to any frame of reference, as we have already discussed. But if $v = c$ in the energy–mass relation given in (14.7), then the energy of the photon $E'_\gamma = M_\gamma c^2/(1 - 1)$ would be infinite – if the photon's mass, M_γ would be non-zero. But the energy of light is certainly not infinite. (When we merely look at something, we are absorbing a stream of photons, converting their energy into signals to our brains, which we interpret as 'vision'). If the photon is a particle that indeed obeys the rules of special relativity theory, its mass, M_γ, must then be equal to zero. This is then a consequence of the result from relativity theory that the speed of light in a vacuum is always equal to the universal speed, c, in all possible inertial frames of reference. It also reflects the fact that, because of the constancy of the speed of light, no external force can speed up a photon or slow it down. All that can happen to a photon is that it can be annihilated – when it is absorbed in ordinary matter.

Coupled with this somewhat peculiar behaviour of the photon, when it is considered as a 'particle', it was discovered in the early decades of the 20th century that under the proper types of experimental conditions, the photon behaves like a bona fide wave, but under other types of experimental conditions, it behaves like a bona fide particle. This discovery of 'wave–particle dualism' in the photon model of light (a discovery made initially by Einstein[2]) had a seminal influence on the future development of atomic theory when it was extended (initially by Louis de Broglie) to the concept of wave–particle dualism for inertial matter as well. This led to the revolution of the quantum theory, that is presently used to explain the features of elementary particles, as the fundamental constituents of matter. [It should be mentioned here that de Broglie's original interpretation of the wave–particle dualism of matter is not the same as the interpretation that evolved in quantum mechanics, whereby the wave aspect of matter has to do with the elementarity of measurement and probability calculus in the laws of nature – according to the prime investigators of this view, Niels Bohr and Werner Heisenberg.]

Question I am not clear about the difference between 'intrinsic' and 'extrinsic' properties of matter, according to the theory of relativity. For example, if a quantity of matter should travel away from a stationary observer at the speed of light, then he should determine that this matter has infinite energy and momentum. If an observer who moves with this matter should measure these properties he would see that its energy is equal to its rest energy, Mc^2 and its momentum is equal to zero. But a quantity of energy is not an illusion! It seems to me that an entity that one can utilize in the forms of electrical energy, heat energy, etc. is something tangible and absolute! Is energy, then, something that one can talk about as an 'intrinsic' property of matter, sometimes, and an entity that is only relative to an observer who is in motion with respect to this matter, at other times?

Reply It is indeed important to make the distinction between extrinsic and the intrinsic properties of matter – extrinsic referring to the measured properties that are due, in part, to the fact that the matter is observed from a reference frame that is in motion relative to it, while intrinsic refers to the properties of this matter that are determined in its own rest frame of reference.

For example, the momentum transfer from a swiftly moving elementary particle, say a proton, to a stationary target, say some heavy nucleus – which is an extrinsic feature of the proton – becomes indefinitely large as the proton's speed approaches the speed of light, relative to the target. Recall that it is the rate of momentum transfer that defines the force that this proton would exert on the target. In a nuclear accelerator (an atom smasher, such as a 'cyclotron'), which is a primary experimental tool of 'high energy' physics, a beam of elementary particles, such as protons, is accelerated to speeds that approach the speed of light. The beam is then focused on other nuclear matter, which are the targets. When the momentum transfer to the target is great enough in a short enough time, the forces that bind the particles of the atomic nucleus target (or that bind an elementary particle to its meson cloud cover), may be overcome, releasing the nuclear fragments (or the created mesons) and then scattered in all directions. It is then anticipated that the distribution of scattered particles, as detected in Geiger counters or bubble chambers, will give the nuclear scientist a clue as to the nature of the nuclear force that was overcome in the process. Here, use was made of the 'extrinsic' features of the matter in motion, relative to a target, to eventually deduce 'intrinsic' physical features and binding of the closed system of particles.

While the scattering experiments in nuclear physics entail the transfer of large quantities of energy and momentum, the intrinsic features of the matter that is interacting, such as the inertial mass, M, or its rest energy Mc^2, are still fundamental properties of matter that should be accounted for by an underlying theory, supposedly being independent of any coupling to other matter. However, when one fully exploits the implications of the theory of relativity, it is found that these intrinsic properties are, in the theory, approximations (expressed in terms of averaged properties) for features that only involve coupling. With the latter view, the inertial mass of matter, M, is not at all an intrinsic feature of matter, as it appears in the theory of special relativity. It is, rather, a particular manifestation of the entire closed system of matter. This follows when special relativity is generalized to general relativity, as we will discuss in Chapter 16, resulting in the application of the Mach principle – a definition of inertial mass of a quantity of matter in terms of its coupling to the rest of a single, closed system – in principle, the universe.

It is also found, with this generalization from special relativity to the general case, that the laws of energy conservation and momentum conservation are only approximations in the 'local domain' (the immediate neighbourhood of the observer), for general relationships that one cannot interpret as conservation laws! The appearance of inertial mass and conservation laws, as intrinsic features of matter, follows from approximating the general form of the theory of relativity (i.e., general relativity theory) to the form of special relativity theory. The difference, as we will discuss in Chapter 18, is that the distinguishing feature of frames of reference in special relativity is constant rectilinear relative motion, while the frames of reference in general relativity in which we compare the form of the laws of nature, are arbitrary types of relative motion.

Thus, so long as one is using the approximation of special relativity (where the accelerated motion is replaced by a series of connected uniform motions) one can separate the extrinsic from the intrinsic features of matter. But when one considers that special relativity is indeed an approximation for general relativity, where in principle a closed system is all that there really is, then the demarcation between 'extrinsic' and 'intrinsic' disappears – as does the atomistic model of matter! These conclusions follow logically then the Mach principle[3] is fully incorporated with the theory of general relativity.

Question From their observations of scattered particles in the atom smashers, do the physicists yet fully understand the nature of the nuclear force, that binds the nuclear particles in the nucleus of an atom? I mean, is there a set of laws here that are analogous to the laws of electromagnetism – the Maxwell field laws?

Reply To date, we have not yet discovered a precise formulation for nuclear forces, in the form of a set of field laws, similar to Maxwell's equations for the electromagnetic forces. Of course, we have learned a great deal about particular features of nuclear forces from these nuclear experiments. One of the important features in the short range of this type of force. For example, the nuclear binding energy of two protons inside of a nucleus is a much stronger attractive force than their mutual repulsive electromagnetic force in this domain. However, at separations much greater than the size of the nucleus, the nuclear force effectively 'turns off' – and the two protons then only repel each other. While we know more about these features, as well as many more details about nuclear forces, we still do not have any fundamental knowledge about the laws of nature that govern the nuclear (so-called 'strong') forces. This is in spite of the great progress that has been achieved in the field of nuclear technology – i.e. advances in the techniques of utilizing nuclear forces. This situation is analogous to the know-how that a chef may acquire to make a delicious cake but without having very much understanding about the chemical reactions that are going on in the baking process. Such knowledge for the chef, as for the nuclear engineer, is called phenomenological, rather than fundamental.

The problems faced by theoreticians to achieve a more fundamental knowledge of nuclear forces have been enormous. Some physicists have proposed (perhaps in despair!) that we should not expect to derive these effects from first principles. We should rather recognize that a phenomenological view of nuclear physics and elementary particle physics (as well as all of the rest of physics!) is all there is to know!

On the other hand, the full form of the theory of relativity takes an opposite stand in which it is assumed that nuclear forces, as well as any other type of interactions between matter, must be derivable from abstract laws of nature which, in principle, include implications about the entire universe, from the domain of elementary particle physics to that of the structure of the universe as a whole – the subject of cosmology. The implication is that there are not truly separated types of forces in nature. They are rather different manifestations of a universal force field, which appears to be gravitational and nothing else, or electromagnetic and nothing else, or nuclear and nothing else, etc., under correspondingly different sorts of experimental conditions of observation. This is an idea first propounded by Michael Faraday, which led him to the unification of electricity and magnetism, under the proper experimental conditions (having to do with the subjective feature of the observer, called 'motion', relative to the observed). Faraday's philosophy then demanded that all forces should be manifestations of a single, unified force field. It was also an implication of the theory of general relativity, as we will discuss in Chapter 21.

NOTES

1 A. Einstein, Does the inertia of a body depend on its energy content?, in: A. Einstein, H. A. Lorentz, H. Weyl and H. Minkowski, *The Principle of Relativity* (Dover, 1923), p. 69.
2 M. J. Klein, Einstein and wave-particle dualism, *The Natural Philosopher* 3 (Blaisdell, 1964).
3 The implications of the Mach principle in general relativity are shown to follow rigorously in: M. Sachs, *General Relativity and Matter* (Reidel 1982).

15

The continuous field concept in relativity

The debate between *atomism* and *continuity*, as fundamental ingredients in our comprehension of the universe, has been going on since the earliest periods in the Western and Oriental civilizations. In ancient times, one asked the question, should one divide up any quantity of matter into finer and finer parts, would there be a final stopping point where one would reach the 'elementary atom' of matter? Or, would it be possible for this matter to be divided up *ad infinitum* in a continuous manner?

The atomists, who claim that there must be a stopping point, have been more successful in convincing most people of their view. Perhaps this is because the idea of discrete atoms more closely matches the human being's responses by way of the mind's perceptions of the world. We tend to think of our surroundings as a collection of individual things.

The continuous field concept is more abstract – it does not directly match our perceptual responses. Perhaps, though, human beings should not expect scientific truth to reveal itself so simply, so as to yield its fundamental nature directly to our senses, or to our constructed instruments. I suspect that we are made in such a way that we require more subtle processes of discovery – using our percepts only for the purpose of extracting hints; then, in a next stage of comprehension, using our ability to reason in order to deduce the essence of the objective truths in science that are sought. Personally, I believe that this is the correct path toward a more complete comprehension of the universe.

With this approach, it is important to recognize that it is impossible for us to conclude any 'final truth' on the basis of our scientific investigations. I believe this to be the case because the amount of comprehension needed to complete any understanding of physical phenomena is infinite, while we are only finite human beings! In this regard it is interesting to recall Galileo's comment[1],

> There is not a single effect in Nature, not even the least that exists, that the most ingenious theorists can ever arrive at a complete understanding of it.

We can only hope to approach the truth by successively matching more and more of the predictions of abstract theories that pertain to the observed facts of nature. I think that this is one lesson that Faraday taught when he introduced the field concept to oppose Newton's 'action-at-a-distance' concept, in a fundamental explanation of the material universe. For Faraday's initial findings about the unification of electricity and magnetism into the electromagnetic field of force was, according to his logic, only an initial step toward a fully unified field that, in principle, should exhibit all of the manifestations of matter in terms of continuity rather than atomism[2].

An important result discovered by Einstein during the early stages of relativity theory was that there is a fundamental incompatibility in maintaining the atomistic view within a theory that fully exploits the implications of the principle of relativity – which is the axiomatic basis of the theory of relativity. Recall that this principle calls for the invariance of the laws of nature with respect to expressions of these laws in different frames of reference that are distinguishable only by virtue of relative motions. In this regard, there is a tacit assumption that motion refers to continuous change. In classical physics, this is a change of the three spatial coordinates with respect to the time coordinate. In relativity, motion is defined more generally, as the continuous change of any of the four space–time coordinates of one frame of reference with respect to any of the four space–time coordinates of any other frame of reference. This generalization follows because time is no longer an absolute coordinate, and space becomes space–time, as we have discussed previously. Such a generalization contains the classical expressions for motion, since, in a particular frame of reference in which the space–time interval appears as a pure time interval, the rate of change of space–time coordinates of a moving object, with respect to the space–time coordinates of that frame, appears as an ordinary velocity in that frame. In both classical and relativistic mechanics, however, motion, is a continuous entity.

The conservation laws give solutions that we identify with conserved quantities, such as energy, momentum and angular momentum. [We actually observe differences in these quantities rather than the quantities themselves.] The conservation laws follow, in this theory, from the invariance of the forms of all of the laws of nature with respect to continuous changes of the spatial and temporal coordinates. The continuous changes are the translations along the space and time directions and the rotations in the (space–space) and (space–time) planes. Thus, the law of conservation of energy follows from the invariance (objectivity) of the laws of nature with respect to arbitrary, continuous shifts of the origin of the time coordinate axis. Similarly, the conservation of the three components of momentum (i.e. in each of the three spatial directions) follows from the objectivity of the laws of nature with respect to arbitrary, continuous shifts of the origin of the spatial coordinate system. The law of conservation of angular momentum follows from the objectivity of the laws of nature with respect to the rotations in space. The translations and rotations are all of the continuous changes in space and time. [The proof that the laws of conservation relate to the invariance of the laws of nature with respect to continuous changes of the space and time coordinate systems follows from Noether's theorem, in the branch of mathematics called 'variational calculus'[3].]

If a real particle of matter is 'here' – therefore if it is not anywhere else – then how can its intrinsic energy content, its rest energy Mc^2, be defined in a continuous fashion throughout all of space and time? The modern atomic theory (quantum mechanics) avoids this difficulty by postulating that energy can only be emitted or absorbed in discrete quantities – 'quanta'. This is as though there would be a filter placed between bits of interacting matter that only allows discrete quantitites of energy to be transferred between them. That is, according to quantum theory, there can be transfers of energy, E_1, E_2, E_3, . . . , but no energy between any two consecutive values, say between E_2 and E_3, can be transferred. No such quantum postulate is assumed in relativity theory, primarily because continuity is imposed there, in the definition of the continuous changes that distinguish different reference frames. This is in accordance with the symmetry imposed in space–time language by the principle of relativity.

It then follows that one cannot accept the discrete particle model of matter in a fully exploited theory of relativity. One must rather deal here with continuous fields – just as Faraday anticipated a century before Einstein. In a letter that Einstein[4] wrote to David Bohm, in 1952, he said:

> When one is not starting from the correct elementary concepts, if, for example, it is not correct that reality can be described as a continuous field, then all my efforts are futile, even though the constructed laws are the greatest simplicity thinkable.

Within the field theory the continuous variables relate to densities that are defined as continuous entities throughout space and time. To compare the predictions of the field theory with observed properties of matter, one must then sum these continuously varying fields over all of space (called 'integration', in calculus) so as to yield the numbers that are to be compared with the meter readings or any other means of recording measurements in physics experimentation.

We see here an example of the abstract approach, to explain the human being's observations of the world. As we have discussed previously, the empirical view taken by present day atomists tends to associate the data from experimentation with the theory itself – it identifies the descriptive level in science with the explanatory level, as is the view of naive realism.

On the other hand, the abstract approach postulates the existence of a theory that leads, by logical deductions only, to predictions that can then be compared with the data of experimentation. This is the case, in formal logic, of a universal leading to particulars. Should any of these particulars disagree with the facts of nature, the universal – the starting theoretical hypotheses for an explanation – would have to be altered in some way, or discarded altogether, and a new theory sought.

With the continuum view of field theory, the apparent discreteness of matter in the microscopic domain is, in reality, a high field concentration in a particular locality, while the field concentrations in other localities may be so weak that they appear to have zero amplitude. This would be analogous to describing particles like the ripples on the surface of a pond. With the atomistic view, any particular atom is separable from the rest of the system of atoms, while maintaining its individuality. However, in the pond example one cannot remove a ripple and study it on its own, as an individual thing – say, by viewing it under a microscope! Still, one can indeed study the motion of an individual ripple in the pond – measure its energy and momentum, locate the maximum of its amplitude as a function of time, and so on. But this 'ripple' is still an entity of the pond; it is a mode of behaviour of the entire pond that is, in principle, without parts. With the atomistic model, the whole is a sum of parts. With the 'holistic' model, there are no separate parts; rather, there is only a single continuous field that represents a closed physical system.

This point of view to a physical model of the universe leads, in a natural way, to the Mach principle – a view of fundamental importance to the continued development of the theory of relativity and is discussed in Chapter 16.

NOTES

1 Galileo Galilei, *Dialogues Concerning Two Chief World Systems,* Stillman Drake, transl. (California, 1967).
2 M. Sachs, *The Field Concept in Contemporary Science* (Thomas, 1973), Chapter II.
3 This conclusion follows from Noether's theorem in field theory. For a mathematical explication of this theorem, see: C. Lanczos, *The Variational Principles of Mechanics* (Toronto, 1966), 3rd edn, Appendix II.
4 Einstein Archives, Jewish National and University Library, Hebrew University of Jerusalem, Call No. 4 1576: 8–053.

16

The Mach principle

Ernst Mach was a philosopher-scientist whose writings, in the later years of the 19th century and the early years of the 20th century, had a profound influence on Einstein's development of the theory of relativity. Certain aspects of Mach's philosophy of science were at first accepted, but later rejected by Einstein. Other aspects of Mach's scholarship had a strong influence throughout the entire development of relativity theory. In addition, and to my mind, of equal importance, was Mach's critical and sharp analysis of existing theories and ideas in physics and the philosophy of science, having a very beneficial influence on Einstein's scientific methodology.

On the method of science, Mach[1] made the following comment.

> The history of science teaches that the subjective, scientific philosophies of individuals are constantly being corrected and obscured, and in the philosophy of constructive image of the universe which humanity gradually adopts, only the very strongest features of the thoughts of the greatest men are, after some lapse of time, recognizable. It is merely incumbent on the individual to outline as distinctly as possible the main features of his own view of the world.

In his comment that 'the very strongest features of the thoughts of men are after some lapse of time recognizable', I would contend that Mach is in agreement with the main thrust of Russell's philosophy of science – implying the existence of a real world, and the possibility of gradually approaching what it is that is true about this world, by a method of successive approximations of theoretical development of our comprehension. Mach's comment also reveals his openness about rejecting ideas about the world, if they can be found to be technically lacking or to have been scientifically refuted.

The latter is not the purely subjectivist view, that many attribute to Mach. From this quotation, we also see Mach's anti-dogmatic approach to scientific methodology. It is thus ironic that Mach's views have been rejected, totally, as dogmatic, by so many well-known scientists and philosophers. The reason for this is, in part, a rejection of Mach's positivistic philosophy of science.

Similar to Berkeley's idealism, Mach supported the view that the only reality that one may talk about in a meaningful way must directly relate to the human being's sense perceptions. Einstein was indeed influenced by this approach in his initial studies of special relativity theory. However, as we have discussed previously, he soon learned that this philosophical approach is not at all compatible with the stand of realism, that was to emerge from fully exploiting the principle of relativity. Rather than his positivistic philosophy, Mach's early influences on Einstein had more to do with his intellectually refreshing criticisms – mainly in his treatise, *The Science of Mechanics*[1]. Here, Mach presented a clear treatment of Newton's physical approach and a criticism of some of Newton's axioms of classical physics. Einstein agreed with his anti-dogmatic attitude – the idea that nothing in science should ever be accepted as an a priori truth. As it has been emphasized in more recent years by K. R. Popper[2], no 'truth' can be established once and for all, as an absolute element of objective knowledge. All that can be proven, scientifically, is a refutation of a scientific assertion, by conclusively demonstrating a logical and/or experimental incompatibility between implications of theoretical bases for natural phenomena and the full set of observational facts.

In *The Science of Mechanics*, Mach[1] criticized Newton's attempts to establish the idea of an absolute space and time. He showed that equally one could derive the classical equations of motion of things from the relativistic view of space.

Mach also offered a very interesting criticism of Newton regarding the manifestation of moving matter, called 'inertia', i.e. the resistance with which matter opposes a change in its state of motion (of constant speed or rest, relative to the reference frame of the source of the force that causes such change). That is, it is more difficult for a tug boat to pull an ocean liner from the dock than to pull a sail boat, because the ocean liner has more 'inertial' mass than the sail boat. In another example, if a large moving van, travelling at only 3 mph, should collide with a brick wall, it would probably disintegrate the wall. But if a man, walking at the same speed, should collide with the same wall, he would probably be thrown to the ground. The difference in these situations is that the greater inertia of the van meant it had much more resistance to the action of the brick wall in stopping its motion than did the man, because of his much smaller inertial mass.

The concept of inertia, as an intrinsic feature of moving matter, was first considered in ancient Greece, and referred to (by Aristotle) as 'impetus'. But its correct quantitative features were not revealed until Galileo's discovery of his principle of inertia. This is the idea that a body at rest, or in constant rectilinear motion, relative to a stationary observer, would continue in this state of constant motion (or rest) forever, unless it should be compelled to change this state of motion by some external influence.

Newton quantified Galileo's principle of inertia with an explicit definition of this 'external influence' or 'force'. Newton asserted that if F_1 is the magnitude of an external force, acting on a given body in causing it to accelerate at the rate, a_1, and if the force F_2 should cause the same body to accelerate at the rate a_2, then the ratio of the forces and ratio of 'caused' accelerations must be linearly related – whatever type of force that is used, i.e.

$$\frac{F_2}{F_1} = \frac{a_2}{a_1} \tag{16.1}$$

(This is an expression of Newton's second law of motion.)

Note also that, at this stage, there is no a priori reason, based on the statements of Galileo's principle of inertia, that this should be a linear relation. That is, Galileo's principle of inertia would not prohibit the possibility that this would be, e.g., a cubic relation, $F_2/F_1 = (a_2/a_1)^3$. But the empirical facts, regarding all of the observed forces in the classical period, including the electromagnetic forces, supported Newton's contention of linearity, as expressed in his second law of motion.

An equivalent way to express the linear relation between force and acceleration is in the usual form of Newton's second law of motion[3].

$$F = ma \tag{16.2}$$

The intrinsic property of matter, called 'inertial mass' and symbolized by the letter m, was then taken to be the constant of proportionality between the external force (the cause) and the resulting non-uniform motion (the effect) of the body acted upon, that is, its produced acceleration. The constant, m, cancels out when this law is expressed in the form of the ratio of two forces that act on the same body.

Mach noticed that the same empirical relation, $F = ma$, could be derived from a different conceptual interpretation of the inertial mass of matter. He argued as follows[1]. Suppose that two different bodies, with masses m_1 and m_2, are accelerated at the same rate by forces with magnitudes, F_1 and F_2. [An important example of this case is the acceleration of different bodies in the Earth's gravitational field.] In this case, the acceleration cancels in the ratio, and the ratio of forces is:

$$\frac{F_2}{F_1} = \frac{m_2}{m_1} \tag{16.3}$$

As in Newton's analysis, this relation may be re-expressed in the linear form

$$m = kF \tag{16.4}$$

where k (= m_2/F_2, say) may be used as a standard for comparison with measures of the inertial masses of all other matter.

Equation (16.4) may then be interpreted to say that the inertial mass, m, of a bit of matter is linearly proportional to the external force, F, that acts on it. Recall that, according to Newton's own definition, F is the total external force that acts on the matter (with mass m), whose source is all other matter of a closed system of matter. Since the gravitational force, for example, has an infinite range, the actual physical system must, in principle, be taken as the entire universe. With this view, taken from a new interpretation of the empirically verified law of motion, $F = ma$, the inertial mass of any quantity of matter is *not* one of its intrinsic features. Rather, it is a measure of the coupling between this bit of matter and all other matter, of which it is an inseparable component.

This interpretation of the inertial mass, which was named by Einstein, 'the Mach principle', may be thought to be analogous to the resistance to the motion of a pellet of copper in a viscous medium, such as oil. This resistance is clearly dependent on the interaction between the metal pellet and the oil. If the oil (and all of the atmosphere) through which the pellet travels, would not be there, and if it would not be gravitationally affected by Earth or any other planet or star, then there would not be any resistance to the motion of this projectile.

One might argue that (16.4) is nothing more than a manipulation of formulae. But so is (16.2)! The actual observables are represented in (16.1) for equal masses accelerated at different rates, or by (16.3) for different masses accelerated at the same rate. It is the ratio formula that relates directly to the data. From such empirical agreements with the ratio formulae, one could then speculate (with Newton) that inertial mass is a measure of a particular intrinsic property of a bit of matter, introduced theoretically as a constant of proportionality between a cause (the applied force, F) and the effect (the caused acceleration a) of the moving matter.

But one can equally speculate (with Mach), from the same empirical data, that the conceptual view of 'inertial mass', as a measure of dynamical coupling within a closed material system, is equally valid. With the latter view, inertia is not an intrinsic property of a bit of matter; it is rather a feature of an entire closed system of interacting material components.

Since all of the manifestations of matter are identified with the effects of forces, and the range of forces (at least the gravitational and the electromagnetic forces) is infinite in extent, the implications of Mach's argument is that any physical system must be considered to be closed. The total mass of the closed system is then not a simple sum of the masses of individual parts – since, with this view, each component mass is a function of the total force that acts where this mass is located. Thus, the dynamical states of motion of all other masses that make up the closed system affects the motion of any component mass of the physical system.

The interpretation of mass, according to Mach – the Mach principle – had a substantial influence on Einstein's development of the theory of relativity, especially its extension from 'special' relativity to 'general' relativity.

At this point it might be asked: Why should only the inertial manifestation of interacting matter depend on the mutual coupling of all of the components of a closed system? What about other features of matter that the atomic theories consider to be intrinsic, such as electric charge, nuclear magnetic moment, and so on? In my own research programme, I have argued that when the philosophical basis of the theory of relativity is fully exploited, it follows logically that this must indeed be the case – that one can no longer talk about 'free particles' at all, as distinguishable, separable parts. Rather, it follows here that one must start at the outset with a single closed system, non-atomistically. All of the 'particle' features of matter – its intrinsic features – must then be derivable, rather than postulated, from the general theory, when certain approximations for the mathematical expression of the general theory can be made. I have called this view[4] 'the generalized Mach principle'.

The mathematical structure of a theory of matter that incorporates the generalized Mach principle, is different from that of the present day theories of elementary particles. In contrast with the linear atomistic approaches of the current theories of matter, such as classical mechanics or the quantum theory, this is a continuum field theory, necessarily in terms of a non-linear structure. In the former scheme, any number of solutions of the original laws form a new, other possible solution. This is called the 'principle of linear superposition'. However, for a closed system, where the mathematical structure is non-linear, there can be no linear superposition.

It is interesting to note that two different aspects of Mach's philosophy have had opposite influences in 20th century physics. The philosophical basis of the present-day interpretation of quantum mechanics – to explain atomic phenomena and elementary particle physics, in accordance with the view of Niels Bohr – is quite close to Mach's positivistic approach. According to the latter view, all that can be said about the atoms of matter must necessarily be limited to the particular reactions of a large-sized (macroscopic) measuring apparatus to physical properties of micromatter. It is basic in this view that, in principle, there is no underlying dynamical coupling that relates the observer (the macroapparatus) to the observed (the micromatter that is measured) in any exact sense. The observer (macroapparatus) is then said to respond to the different states of the observed (the micromatter) with various degrees of probability. The laws of nature, in this view, become laws of probability – a 'probability calculus'. This theory of micromatter is then fundamentally non-deterministic and it is based on the particularistic notion of atoms with imprecise trajectories, that is, without predetermined specifications of all of the parameters that characterize these trajectories simultaneously, with arbitrarily precise values.

This positivistic, non-deterministic view is in constrast with a different aspect of Mach's philosophy – the Mach principle – which implies that any material system must be closed at the outset. The theory of relativity also implies this view. As with the quantum theory, one must start out with an 'observer' and the 'observed' to make any meaningful statement about a physical system. But in contrast with the quantum approach, this is a closed system, holistically described, without actual separable parts. 'Observer' is not a defined concept without the existence of the 'observed' – just as the concept 'early' is meaningless without the concept 'late'. Further, and of equally important distinction, there is no intention in the theory of relativity to define 'observer' in anthropomorphic terms, or in terms of large scale quantitites with respect to small scaled 'observed' matter. In this theory, observer–observed is taken to be a closed system. The hyphen is only used for purposes of convenience in particular situations where one can approximate the closed system by one component that is very weakly coupled to the rest of the system – sufficiently so that, under these circumstances, one can treat one of the coupled components to be an almost 'free' quantity of matter, and the other, the 'apparatus' that responds to this body. But it is important with this view that, both from the philosophical and the mathematical standpoints, there is, in reality, no 'free matter', and no separated apparatus that looks down on it in a purely objective manner. The only objective entity, in this view, is a single, closed system, not composed of separate parts. This is the universe.

Question How do you feel the implications of a closed material system, according to the Mach principle, may relate to the human society?

Reply With the philosophy of science that takes any realistic system to be closed at the outset, there seems to me to be an overlap with particular views in sociology and psychology. Many physicists do not accept this holistic approach – indeed, the majority of contemporary physicists are atomists who adhere to the notion that any physical system is an open set of things, even though, according to the present-day view of quantum mechanics, the individual things are not said to have predetermined physical properties. Similarly, I'm not sure that many of the contemporary sociologists and psychologists accept the view of a closed system underlying the human society. Recall that 'closed system' does not refer to a sum of parts, such as a collection of human beings who are free of each other, except for their interactions. It rather entails the society as a single entity that is without actual parts – even though it manifests itself as a system of weakly coupled parts, under the proper conditions. It is nevertheless important in the descriptions of both the material, inanimate physical system, studied by the physical scientist, and the human society, studied by the social scientist, that the closed system does not predict observable features of the whole that are strictly a consequence of the intrinsic properties of 'parts' – it is rather a single entity without any actual parts.

It seems to me that it would take a great deal of sacrifice for any human being to accept and truly believe this idea about the elementarity of the closed system. It is an idea that seems to force the individual to yield the supremacy of his or her ego to the whole of society. That is, a person's ego tries to convince him or her that they are free individuals, apart from the rest of society – certainly interacting with society, but according to the whims of their own free wills. This is opposite to the view of a person as a particular manifestation of the closed system, where 'free will' is only illusory, resulting from the individual's state of partial ignorance. With the latter view, the human being's actions are predetermined, according to the mutual interactions of the entire, closed society. Thus, for the human being to act in one way or another is, in this approach, predetermined, and indeed imposes restraints throughout the entire system, similar to the effect on a tightly inflated balloon due to pressing one's thumb into it, anywhere!

NOTES

1 E. Mach, *Science of Mechanics* (Open Court, 1960).
2 K. R. Popper, *Objective Knowledge: An Evolutionary Approach* (Oxford, 1972). In this text, see the topics listed in the Index under the title 'dogmatism'.
3 The original expression of Newton's three laws of motion is in his treatise, I. Newton, *Principia* (California, 1962), Vol I, A. Motte, trans., F. Cajori, revised.
4 I have developed this idea in: M. Sachs, *General Relativity and Matter* (Reidel, 1982).

17

Experimental confirmations of special relativity and transition to general relativity

The combination of

- *the principle of relativity* – saying that the laws of nature must be independent of any frame of reference in which they may be expressed, by any particular observer, and the implication that they must be in terms of continuous field variables, and
- *Mach's principle* – implying that any real physical system in nature must be closed

leads to the theory of general relativity. The starting point of this generalization was Einstein's question: Why should the principle of relativity be confined only to comparisons of the laws of nature in inertial frames of reference, that is, frames that are in constant rectilinear relative motion? Should one not expect that the laws of nature would be the same in frames of reference that are in arbitrary types of relative motion? His answer was that this must indeed be the case – since *all* motion is relative – as Galileo taught us in the 17th century! That is to say, if there should be a law of nature implying that Tamar accelerates relative to Jose, because there is some physical cause of her affected motion, then there must be an identical law, leading to the conclusion that Jose is accelerating relative to Tamar, as predicted by a corresponding cause–effect relation, but from Tamar's frame of reference rather than that of Jose.

This is meant in the same way as we discussed previously. If Tamar should deduce laws of nature from her observations of physical phenomena, within her own space–time reference frame, and if she should deduce the form of the laws of nature in Jose's reference frame, in terms of the space and time coordinates of that frame relative to her own, then she should find that the two expressions of a law of nature for any physical phenomenon in nature are in one-to-one correspondence, if these are indeed bona fide, objective laws of nature.

We note that non-uniform motion is the only type of motion that can actually be experienced by matter, when it interacts with other matter. This is because when one bit of matter reacts to other matter, this is due to a force that acts on it (by definition), causing a transfer of energy and momentum to it from the other matter. Force, *per se*, is the cause of non-uniform motion. Thus, the case of uniform motion, which underlies special relativity theory, is an extrapolation from interaction to the ideal case, where the interaction turns off! That is to say, 'special' relativity is a limiting case that is, in principle, unreachable from the actual case of general relativity – where the only kind of relative motion between reference frames in which one compares the laws of nature is non-uniform.

Even so, one knows from the empirical evidence that the formulae of special relativity theory work perfectly well in describing a host of different sorts of experimental situations. To name a few of the important ones, it explained the results of

- the Michelson–Morley experiment,
- the Doppler effect for electromagnetic radiation,
- changes in the momentum and energy of matter, as it moves at speeds close to the speed of light, that are different from the predictions of classical mechanics, and
- the well-known energy-mass relation, $E = Mc^2$, which has been well-verified in nuclear physics experimentation.

Still, there are properties of matter, observed in domains where special relativity provides an adequate description for other properties, that have not been explained within the formal expression of special relativity. An example is the set of physical properties associated with the inertia of matter, such as the magnitudes of the masses of the elementary particles – electron, proton, pi meson, and so on – and the empirical fact that they lie in a discrete spectrum of values. Also not explained within special relativity is the feature of the gravitational force that it is only attractive, in the domain where Newton's theory of universal gravitation had been successful. This implies that the inertial mass of any bit of matter has only one polarity. This is in contrast with the electric charge of matter, that has two possible polarities, implying that electrical and magnetic forces can be either attractive or repulsive.

Let us now review some of the experimental data that have substantiated the formal expression of the theory of special relativity. The first is the *Michelson–Morley experiment*. This was carried out in the last years of the 19th century. Its original aim was to measure the speed of the aether drift, relative to the Earth. (Everyone in that day believed aether to be the medium to conduct the flow of light.) The idea of this experiment was that as the Earth spins on its axis, it must be in motion relative to the stationary, all-pervasive aether. As we observe the aether from a position on the spinning Earth, we should detect the effect of the aether on the speed of the propagating light waves, as they move in different directions relative to the Earth's axis. It was expected that the speed of light should be different in the directions parallel to the Earth's surface and perpendicular to it. This is analogous to the motion of a small boat relative to a flowing river. If the river flows at V cm/sec in the eastwardly direction, then if the boat's speed in still water would be v cm/sec, Jill, who stands by the shore of the river would see that if the boat travelled eastwardly, its speed relative to her would be $v + V$ cm/sec; if it travelled westwardly in the river, its speed would be $v - V$ cm/sec relative to her, and if it crossed the stream in the northerly direction, its speed relative to the shore would be $(v^2 + V^2)^{\frac{1}{2}}$ cm/sec. In the Michelson–Morley experiment, v is the speed of the propagating light, if it should be in a vacuum, and V is the speed of the aether, relative to the Earth's surface, which is equivalent to the speed of the Earth relative to a stationary aether.

Consider a beam of light with a single frequency (called monochromatic), say the yellow light from a sodium lamp, as it is split along paths of equal lengths, but at different speeds. If the two beams should be brought back to the starting point with the use of mirrors, one should detect a loss of synchronization of their phases when they rejoin, if they were initially synchronized in phase. Of course, this is because it would take different times for the two light beams to traverse the same distances, if they travel at different speeds.

In the Michelson–Morley experiment, where one uses an 'interferometer' to study the comparison of the phases of the light beams, one having moved away from a source and back to it in a direction parallel to the surface of the earth, and the other moving away from the source and back to it in the direction perpendicular to the Earth's surface, one should then expect the recombined waves at the location of the source, because they are then out of phase, to yield an interference pattern. From this interference pattern one could then deduce the speed of the aether relative to the Earth's surface – with very high accuracy. Beside the magnitude of this speed, such an experiment would prove the existence of the aether to conduct light.

The result of this experiment[1] was a null effect! That is, there was no interference pattern observed. One possible implication of this result was that there is no aether in the first place! It would be analogous to a vehicle travelling at a constant speed in a dry river bed – its speed seen to be the same in all directions – because there is no river there to alter it. If the observer could not see whether or not there is a flowing river, but he could observe that the vehicle travels at the same speed in the river bed in all directions, he would probably conclude that there is no river in the bed where it was supposed to be. In this way, the only reasonable conclusion from the Michelson–Morley experiment was that there was no aether there to conduct light.

Such a resolution of the negative result from this experiment was inconceivable to most scientists of the late 19th and early 20th centuries – including Michelson himself. He firmly believed (with Maxwell, had he lived to see their experiment) that the radiation solutions of Maxwell's equations are an expression of the phenomenon of light as the vibrations of an aether – analogous to the interpretation of sound as the vibrations of a material medium.

Independently, Lorentz (in Holland) and Fitzgerald (in Ireland), inter-preted the negative result of the Michelson–Morley experiment as a physical property of the aether, in that it interacts with physical instruments in such a way that they must contract by different amounts, depending on their direction of motion in the aether[2]. Their formula for the amount of physical contraction, due to the force exerted by the aether on the measuring instruments (in the direction of its motion), was precisely the 'Lorentz transformation', discussed earlier in Chapters 10 and 11.

Einstein's theory of special relativity, which was published several years after the Michelson–Morley experiment, though not referring directly to it, did explain their negative result in a logically consistent, theoretical fashion. It was an explanation in terms of the idea that the aether is indeed a superfluous concept in physics – that it need not exist to explain the propagation of light. The explanation here was in terms of the relativity of space and time measures, that in turn led to the Lorentz transformations. As we have discussed earlier, Einstein's explanation was along the lines of treating these transformations as scale changes of the space and time measures, when an observer expresses a law of nature in a frame of reference that is in motion relative to his own. Thus they are no more than the 'translations' from the language appropriate to one reference frame to that of another, in order to preserve the form of the law (in accordance with the requirement of the principle of relativity.)

It is interesting to note that there is still controversy among historians of science on the question about whether or not Einstein was aware of the Michelson–Morley experimental result when he formulated special relativity theory. It is also interesting that Michelson, himself, was not too happy about the explanation of his result with the theory of relativity – because of a feeling he had that this theory could not be correct! This points to advice that when analysing the history of science, one must take account of the fact that, far from being totally objective thinking machines, scientists are only human – along with all of the 'hang-ups' that go with this label! – such as irrational prejudices in science to fight off, as well as other emotional restraints, such as the near-omniscience that the scientist sometimes attributes to the leaders in his field!

The Doppler effect, in regard to the propagation of light, has a precedent in the classical description of wave motion. For example, when an ambulance or fire-engine passes, while sounding its siren, the sound heard has an increasing pitch (frequency), as it approaches, and a decreasing pitch as it departs; this phenomenon is called the Doppler effect, and was discovered in regard to the propagation of sound by Christian Doppler, in the 19th century.

In a more quantitative investigation of this effect, it is found that the listener should move relative to the source of the sound *or*, if the source of the sound should move relative to the listener, in the opposite direction but with the same speed, the measured frequency shift would be slightly different in each of these cases. The reason is that in the second case mentioned, where the source of sound is in motion relative to the listener, the oscillations of the density of the air molecules (which accounts for the phenomenon of sound) will 'bunch' the conducting, vibrating medium in the direction of motion of the source of the sound, while no such bunching effect occurs when it is the listener who is in motion relative to the source of the sound. Thus, the different Doppler effects for each of these cases is due to the existence of a medium whose role is to conduct the sound, (by compressing and rarefying, in time, to create the 'sound waves').

In the Doppler Effect it is the frequency of the measured sound wave that is changing. The quantitative effect that is predicted, when the listener moves away from the sound source, at v cm/sec, is:

$$f_m = f_s\left(1 - \frac{v}{c'}\right)$$

where f_m is the frequency of the sound (measured in cycles per second by the listener), f_s is the actual (proper) frequency of the sound emitted by the source and c' is the speed of sound in this particular medium.

In the second case, where it is the source of sound that is in motion relative to the listener, (in the opposite direction, but with the same magnitude of relative speed), the measured frequency is:

$$f_m = \frac{f_s}{\left(1 + \frac{v}{c'}\right)} \text{ cycles per second}$$

According to the binomial expansion, $1/(1 + v/c')$ is practically the same as $(1 - v/c')$, if the ratio v/c' is small compared with unity. (The actual difference depends on an infinite series of terms, in increasing powers of this ratio, starting with $(v/c')^2$.) Thus, if the latter infinite series of terms can be neglected, these two Doppler effects would be numerically equal to each other. Nevertheless, these two effects are really measurably different, even if by a small amount when v/c' is small compared with unity (the usual case)[3].

The Doppler effect in relativity theory, for radiation, is independent of whether the source of that radiation moves relative to the observer of it or vice versa. This is because the time coordinate is no longer an absolute measure – it is a subjective parameter, depending on the frame of reference in which it is expressed. Further, as we pointed out before, there is no medium (aether in this case) to conduct light – thus, there is no conducting substance to 'bunch' in this case, as happens with the propagation of sound[4].

The numerical predictions for the exact amount of frequency shift, according to the Doppler effect in relativity theory, has been experimentally confirmed. Further, the theory predicts a Doppler effect when the frequency of light is measured in a direction that is perpendicular to the direction of the propagation of the light. This is called the 'transverse Doppler effect'. This effect has been experimentally confirmed in the case of electromagnetic radiation; it has no classical counterpart in the case of sound.

The changes in energy and momentum of matter, as a function of its increasing speed, according to the formulas of special relativity theory rather than classical physics, have been verified in numerous experimental tests. For example, the designs of particle accelerators and the design of the mass spectrograph, are based on the dynamics of swiftly moving particles according to the predictions of the theory of special relativity.

The verification of the energy-mass relation, $E = Mc^2$, has been most outstanding in the observations of nuclear disintegration, such as nuclear fission, whereby heavy, unstable nuclei break up into a number of lighter ones, and nuclear fusion, such as the processes of interactions between light nuclei, giving rise to an emission of large amounts of energy, such as the processes of energy emission occurring in the Sun.

All of these data have been spectacularly supported by the predictions of the theory of special relativity. Still, according to what has been said earlier, this success can only be taken as an indication of the accuracy of the theory of special relativity as an approximation for a formal expression in general relativity. For in spite of the remarkable accuracy of the formulae in special relativity in representing particular high energy data, it does not work in explaining other data in these same domains of measurement. For example, the inertial mass parameter must be inserted into the special relativity formulas, later to be adjusted to the experimental data. On the other hand, according to the full expression of the theory of general relativity, the mass parameter must be derivable from the general physical features of the closed system. The general theory that underlies such a derivation will be discussed in Chapter 18.

Question Is it possible that the theory of special relativity is no more than an accurate way of describing rapidly moving, electrically charged particles and electromagnetic radiation, while the theory of general relativity is an entirely different theory – a theory that has to do with gravitational forces and nothing more?

Reply There are critics of the general conceptual structure of relativity theory who argue that, because the formulae of the theory of special relativity 'work', there is no real justification for further generalization. They say that the particular application of the mathematical apparatus of general relativity theory to describe successfully planetary motion and a few other phenomena having to do with gravitational forces, shows that this is an accurate description of gravitation, but it is not an explanation – indeed, that there is no need for further explanation!

This group of physicists take the philosophic stand of logical positivism. That is, they look upon the formulae as no more than an economic way to describe the data, but they deny that they can come from any more basic ideas, such as an abstract law of nature that plays the role of a universal that underlies the empirical facts. They claim that if some investigators have been lucky enough to derive predictions of the data from such claimed 'underlying universals' then it was purely coincidental (and perhaps a bit of chicanery!)

Of course, one can always take a set of formulae that happen to be 'working' at the time, in being able to fit particular data, and deny that there is anything more to talk about. But I would then claim that these people are at a dead end, as far as scientific progress is concerned! With their point of view, they cannot make any progress in understanding the world – i.e. in deriving some of its features from first principles. All that they can do is to wait until more formulae appear on the scene, as will undoubtedly happen as science progresses. But these changes will not be discovered by the empiricists and positivists – unless a few miracles accidentally happen to drop the right formula in the right place at the right time! The changes in scientific understanding invariably come from investigations that are foundational in regard to the laws of nature. These, in turn, imply, by logical deduction, particulars such as the predictions of special types of data for corresponding experimental arrangements.

In the meanwhile, the empiricists and positivists will guard the existing formulae, as well as the meaning they attribute to them, with their lives! They will be the most fervent opponents to any real changes in the formulae – because, to this group of scientists, the data is the theory! Since the data cannot be wrong, they argue, their theory must be true, absolutely. In this philosophy, it is not usually recognized that whatever one says about the data must be expressed within the language of a particular theory, and this theory could be (in part, or totally) wrong! Indeed, we have seen this to be the case repeatedly throughout the history of science.

An example of the procedure in which one starts with a universal (a general law) and then derives particulars to be compared with the observations, is the theory of general relativity, with special relativity playing the role of a mathematical approximation that applies to some, but not all physical situations. If an implied particular should be found to disagree with a single observed fact, then the entire theory would be challenged –

as though one were to pull a single thread from a woven blanket, thereby causing an unravelling of the entire blanket. But so long as there is agreement between the predictions of the theory and the empirical data, one can say that he or she 'knows' that, thus far, the scientific theory is valid. On the other hand, when one proceeds in the opposite direction, going from particulars to a possible universal (general theory), then I do not see that one can ever say that he or she 'knows' anything, except for the tabulation of data. This is because a given set of data can lead to any number of possible universals. Yet, the latter approach, using the method of inductive logic, is indeed followed by the majority of contemporary physicists, especially those most influenced by Bohr and Heisenberg in this period of physics. The former approach of the theory of general relativity theory, on the other hand, taking the philosophic stand of realism and using the method of deductive logic to draw conclusions about what is and what is not (scientifically) true, seems to be to me a superior method of investigating truth, since it allows one to talk about what he or she 'knows', in addition to the readings of instruments – at least until any of this knowledge may be scientifically refuted.

NOTES

1 See, for example, A. P. French, *Special Relativity* (Norton, 1966), p. 51–7.
2 *Ibid.*, p. 63.
3 Any good elementary text on physics should have a discussion of the Doppler effect for sound waves. See, for example, D. Halliday and R. Resnick, *Fundamentals of Physics* (Wiley, 1988), Sec. 18–7.
4 For a detailed discussion of the relativistic Doppler effect, see: H. M. Schwartz, *Introduction to Special Relativity* (McGraw-Hill, 1968), Sec. 3–3.

18

The curvature of space–time

It was previously pointed out that 'motion' in relativity theory is defined in a more general sense than it is in classical physics. I will now elaborate on this somewhat further. With the classical approach, when we talk about space and time, we refer to absolute coordinates of measure. The classical 'motion' then refers to changes of spatial measures with respect to a time measure. For example, the x_1-component of the velocity of a particle of matter is called $\delta x_1/\delta t$ cm/sec, where δt denotes the same change of time measure to all observers. Thus, $\delta x_1/\delta t$ is the 'rate of change' of the spatial measure in the x_1 direction with respect to the time change δt.

However, according to special relativity theory, the single spatial or temporal coordinate, in a particular frame of reference, becomes a four-dimensional space–time interval in any other (relatively moving) reference frame, where the laws of nature are compared. Then, 'motion' refers to the changes expressed with the parameters $\delta x'_\mu/\delta x_\nu$, where the Greek subscripts refer to the four space–time coordinates ($x_0 = ct$, x_1, x_2, x_3) in one particular reference frame and the primed coordinates, ($x'_0 = ct'$, x'_1, x'_2, x'_3) in a different reference frame that is in motion relative to the first. The primed and unprimed reference frames then refer to the coordinate systems in which the laws of nature are to be compared, by a given observer, say the one whose coordinates are unprimed.

The restriction imposed by the theory of special relativity, is that the rates of change between the coordinates of one reference frame and another, $\delta x'_\mu/\delta x_\nu$ are those transformations from the unprimed to the unprimed reference frames that leave invariant the four-dimensional interval,

$$\delta s^2 = \delta x_0^2 - \delta x_1^2 - \delta x_2^2 - \delta x_3^2 \tag{18.1}$$

$$= \delta x_0'^2 - \delta x_1'^2 - \delta x_2'^2 - \delta x_3'^2$$

A feature of the 'motion' in special relativity theory, as characterized by these rates of change, is that they are independent of the positions in space–time where they are evaluated. There are 10 such rates of change in special relativity theory. These are: three components of constant, rectilinear velocity of one frame relative to the other, $v_i/c = \delta x_i/(c\delta t)$, $i = 1, 2, 3$;

three angles of rotation in the three-dimensional space (called the 'eulerian angles') and four continuous translations along the space and time coordinate axes (shifting the origin of the space and time axes by arbitrary amounts).

Note that the interval δs^2 is also unchanged under reflections of the coordinates, $x_\mu \rightarrow -x_\mu$. These are discrete transformations, therefore they are not generally required by the postulates of the theory of relativity, which depend only on changes that are continuous relative motions in space and time.

Einstein discovered that to describe a general (non-uniform) relative motion, the metric of space–time with $\delta s^2 =$ constant according to (18.1) in special relativity, must be generalized further. The more general geometry of space–time is one in which the continuous changes, $\delta x_\mu'/\delta x_\nu$ are no longer constant in space–time, but rather, their values depend on where in space–time they are evaluated. When these rates of change are independent of the places where they are evaluated they are linear transformations; in the latter case, they are non-linear transformations.

According to Einstein's approach, the mass distribution for a physical system determines the logic of the space–time language (i.e. its geometry), which, in turn, is used to express the laws of physics of any type. Since the mass distribution is a continuous entity, as expressed in terms of 'fields' in space–time, it follows that the geometry of space–time must similarly be characterized by space–time-dependent parameters.

To attain such a generalization of the geometrical part of the logic of space–time, Einstein proceeded from euclidean geometry, that characterizes special relativity, to the non-euclidean geometry discovered by Riemann, for general relativity[1]. In this case, the constant (linear) metric for special relativity, given in (18.1), is replaced by the more complicated, non-linear metric of the Riemannian space–time:

$$\delta s^2 = g^{00}\delta x_0^2 + g^{11}\delta x_1^2 + g^{22}\delta x_2^2 + g^{33}\delta x_3^2 \qquad (18.2)$$

$$+ g^{01}\delta x_0\delta x_1 + \delta^{02}\delta x_0\delta x_2 + g^{03}\delta x_0\delta x_3$$

$$+ g^{12}\delta x_1\delta x_2 + g^{13}\delta x_1\delta x_3 + g^{23}\delta x_2\delta x_3$$

$$= \text{constant}$$

where each of the 10 coefficients, $g^{\mu\nu}$, is a function of where in space–time it is evaluated. This collection of 10 coefficients is called the 'metric tensor'.

From the geometrical point of view, one might look on the constancy of the right-hand side of this invariant interval, in (18.2), as the equation for a 'squashed cone' – it is a non-linear generalization for the 'linear' light cone, discussed in Chapter 13 in regard to the special relativity metric.

According to Riemann's 'differential geometry', the components of the metric tensor have the boundary conditions imposed on them as follows. As the system becomes emptied of matter, so that the reference frames become inertial (special relativity),

$$g^{00} \to 1 \quad g^{11}, g^{22}, g^{33} \to -1 \quad g^{\mu \neq \nu} \to 0$$

In this limit, we see that the general relativity metric, in (18.2), approaches the special relativity metric, in (18.1). This is a requirement of the empirical facts, that the mathematical description in general relativity continuously approaches the mathematical description in special relativity, as the matter of the system is correspondingly diminished.

One of the outstanding features of Riemann's geometry is that it does not contain the axiom of parallel lines (see page 00). In contrast with euclidean geometry, with Riemann's geometry, all 'straight' lines intersect, somewhere.

The term, 'straight line', means 'geodesic' – that particular locus of points that minimizes (or maximizes) the distance between any two of them. Riemann's geodesics in a space–time, where there are no parallel lines, is analogous to the family of great circles of a sphere. In this case, any great circle (a circumference) on the surface of the sphere, that contains a point on the surface of the sphere, outside of a second great circle of the sphere, will intersect the second great circle, somewhere on the surface of the sphere. These are similar to the family of longitudes of the Earth all going through the north and south poles.

An important feature of Riemann's space–time is that its family of geodesics are not straight lines – they are curves. That is to say, the shortest distance between any two points of his space–time, is a curve rather than a straight line. This is analogous, again, to the fact that the shortest distance between New York and London is on a great circle of the Earth – passing through Laborador and other very northern points on the surface of the Earth, rather than the straight line path that one might draw between New York and London on a flat map.

The latter is only an analogy. The 'curvature' of a four-dimensional space–time is not directly perceivable with our senses! For example, if one should look at the surface of a sphere from the outside, it would be seen to be convex – bending away from the observer. This may be defined as 'positive curvature'. But if the surface of the sphere should be viewed from the inside, it would be seen to be concave – bending toward the observer. This would be 'negative curvature'. On the other hand, in a riemannian space the curvature might be positive everywhere or negative everywhere – here, there is no inside or outside of space to talk about! This feature cannot be directly visualized with the human senses. Still, it is perfectly legitimate, in terms of a mathematically rigorous description of space–time, to utilize such a geometry in representing the laws of nature. Such a mathematical description in general relativity theory indeed led to predictions that were tested regarding the experimental facts. This is all that we should require of the mathematical language that expresses a law of nature. The set of all geodesics that characterize the riemannian space–time is a 'curved space–time'.

One of the implications of the 'curvature' of space–time, that leads to the prediction of observable effects, is that the path of a 'freely moving' object in such a space would appear to a 'stationary observer' as a curved path. This would seem to be in contradiction with Galileo's principle of inertia, since the statement of this principle is that any body that is unimpeded will naturally move in a straight line path. In addition, Fermat (who lived in the same period as Galileo) discovered the principle of least time for light, asserting that it must propagate in such a way that the time taken to go from one point in a homogeneous space to another is a minimum – implying that the path of propagating light is a straight line. However, the latter conclusions for light and material particles were based on the assumption of a euclidean geometry, whose geodesics are straight lines. [Of course, Galileo and Fermat, in the 17th century, had no awareness that there could be any geometry other than that of Euclid!] On the other hand, with the non-euclidean geometry of general relativity, the prediction would then follow that a freely moving body or a light beam must travel on a geodesic that is a curve. Such an observation of the curved path of starlight, as it propagates past the rim of the Sun, was made in the early stages of general relativity theory and it was confirmed both qualitatively and quantitatively in experimentation, in agreement with the precise predictions that come from the field equations of Einstein's theory.

The actual equations of motion for the path of a freely moving object follows from a minimization of a general type of path between any two points of a riemannian space–time. This leads to an equation whose solutions reveal the shape of the path of the moving object. When this is done with a euclidean metric, as in special relativity theory, it is found that the geodesic is indeed a straight line, as observed from any frame of reference. The classical result is predicted from an equation of motion that has the form: *acceleration* = 0. The solution of this equation is: *velocity* = *constant* and the solution of the latter equation is:

$$distance\ travelled = velocity \times time + constant$$

The latter equation then predicts that the distance covered by the freely moving body depends linearly on the time. This relation, in turn, yields the straight line path of the body. Thus, in special relativity, as well as newtonian physics, the shortest distance between any two points travelled by a freely moving body – its geodesic – is a straight line. This is a feature of the underlying euclidean geometry of the space and time.

Using the same sort of analyses to determine the shape of a geodesic path in a riemannian space–time, with the invariant metric shown in (18.2), the equation of motion takes the form:

$$acceleration + something\ else = 0$$

The 'something else' in this equation is a complicated mathematical form that depends on the metric tensor field, $g^{\mu\nu}(x)$ and its rates of change with respect to the space and time coordinates – in a form that quantitatively defines the curvature of space–time. If one should now call this 'something else' by the name, $(-F/m)$, then the equation of the freely moving body, in the riemannian space, would have the form of Newton's second law of motion,

$$acceleration = F/m.$$

Recall that, according to Newton's theory of motion, the symbol F refers to a cause for the acceleration of the body. This is, by definition, an external agent, in Newton's theory, acting on a quantity of matter with inertial mass m, causing it to accelerate at the rate a. But in Einstein's analysis, within general relativity, there is no 'external' force. It is rather that we are describing the motion of a 'free' particle of matter in a space–time with a more general type of geometry than the one that was used in Newton's analysis. The initial assumption made, however, is that a free matertial body must move along the path that is, geometrically, a geodesic – the path of minimal separation between any two of its points.

In order to build this feature into the theory, the law of nature that predicts the detailed behaviour of the metric tensor, $g^{\mu\nu}$, (that yields the shape of the geodesic) from the variables of matter, takes the 'source' of the metric tensor to be another variable that relates to the energy and momentum of the material system that is influencing the moving body. In this way, the metric tensor solution that predicts the geodesic path for the observed matter is that solution that also corresponds to the minimum energy path of the moving matter. That is, any non-geodesic path for the moving matter would have to correspond to putting extra energy (work) into the moving body. In this way, the 'matter' variables are related to the geometrical variables in terms of dynamical equations of motion. The law of nature whose solutions are the components of the metric tensor, $g^{\mu\nu}$, as due to the existence of the energy–momentum content of the system, are 'Einstein's field equations'.

With the result that Newton's equations of motion, expressed in terms of his second law, $F = ma$, could be derived from geometrical considerations, Einstein saw the possibility that all external forces may only be apparent – that the effects of other matter on a test body may be representable with nothing more than a generalization of the geometry of space–time with which we describe the motions. Following this idea through, he then showed that the force of gravity (at least) can be fully represented and explained in terms of the motions of free things in a space–time with the generalized riemannian geometry. The idea then persisted in his thinking that perhaps further generalization of the geometrical logic of space–time might lead to all other force manifestations of matter, in a unified field theory. This approach provided Einstein with a more logically satisfying explanation for interacting matter than Newton's theory of universal gravitation, in understanding gravity. This idea will be discussed more fully in Chapter 21.

Question Did Einstein's plan to geometrize matter mean that he wished to introduce the effect of one massive body on another massive body by thinking that the former would 'warp' the space–time in the vicinity of the latter and thereby alter its path?

Reply To explain this idea of geometrizing matter, according to Einstein's plan, think of an external force acting on some 'particle of matter' – say the effect of the Sun (in terms of its gravitational potential) acting on the speck of a planet that we call Earth. Classically, one thinks of this description in a flat (euclidean) space and time. But, alternatively, one may now think of this same description of the effect of the Sun on the planet (in causing its orbital motion) by removing the Sun from the picture, but replacing it with a 'free motion' of the Earth in a particular sort of curved space–time, rather than its 'free motion' in euclidean space–time, to yield its motion along a curve rather than a straight line path. This generalization is not an effect of the Sun; it is, rather, a way of representing the Sun's existence in a geometrical fashion!

With this approach, Einstein showed, from the solutions of his field equations (which explicitly relate geometrical variables to matter variables) that in addition to duplicating all of the successful results of Newton's theory of universal gravitation, he was able to make further predictions that were substantiated by the experimental facts, but not even predicted in qualitative fashion by the classical theory. In this way, Einstein's theory had indeed unseated the newtonian theory of universal gravitation[2].

Question If the space–time of the universe is curved and closed, does this imply that a trajectory of light, when sent into outer space, would come back to its starting point in space?

Reply Yes, but one would not be able to wait long enough to see this happen! That is, just as a ship that heads away from the dock, travelling on a great circle of the Earth, will return to the dock, so a beam of light that is propagated into outer space should traverse a 'great cosmic circle', eventually returning to the starting point. This assumes, of course, that this light would not be absorbed by other matter along the way. This conclusion is consistent with the feature of general relativity theory, when it is incorporated with the Mach principle, that the universe is a closed system. [Note that in this comparison, the ship is travelling along a two-dimensional surface of a three-dimensional sphere, while the light beam travelling into the cosmos in a four-dimensional rienannian space–time is moving along a three dimensional surface of a four dimensional space–time continuum. Our senses can visualize the former description in the euclidean space but they cannot visualize the latter description in the reimannian space–time. Nevertheless, this is a mathematical formulation that does lead to predictions that we can, in principle, detect with our human senses or measuring devices – such as the absorption of the light signal after it had traversed a cosmic great circle, if we could wait long enough!]

NOTES

1 A clear discussion of non-euclidean geometry and its role in general relativity, for a non-technical audience, is in: C. Lanczos, *Albert Einstein and the Cosmic World Order* (Wiley, 1965).
2 A detailed analysis of Einstein's theory of general relativity and its prediction of gravitation is in: M. Sachs, *General Relativity and Matter* (Reidel, 1982).

19

Gravitation and crucial tests of general relativity

In Chapter 18, we discussed the idea that a free particle of matter in a riemannian space–time would appear to any observer to move on a curved path, as viewed from their own frame of reference – as though it had been subjected to an external force. The actual shape of this path would depend on the specific dependence of the metric tensor, $g^{\mu\nu}$, on space–time. This dependence, in turn, enters the definition of the invariant metric in general relativity, δs^2, as shown in (18.2), in Chapter 18.

A question that now arises is the following: Where does this set of 10 field variables, that make up the metric tensor $g^{\mu\nu}$, come from, in the first place? This was the question that Einstein addressed when he initiated his studies of general relativity theory. For once the correct field equation could be found, whose solutions are the metric tensor variables, then these may be inserted into the geodesic equation (into the term that we previously called 'something else') – that serves as the equation of motion of a test body. The solutions of this equation would then predict the path of a test body – be it the orbit of Earth or the other planets about the Sun (or vice versa, according to Galileo's observation about the subjectivity of 'motion'). [That the geodesic equation could serve as the equation of motion is based on the assumption that the moving body itself has minimal energy. But there is an additional assumption here that the contribution to the total geometrical field due to the presence of the test body itself is of negligible importance – which is sometimes a good assumption, as with the influence of the Earth's own field in the total field of the Sun, and sometimes it is not a good assumption, as would be the case in the analysis of a double star system.]

To guide him in his discovery of his field equations, that could at least serve as a preliminary test of the theory, Einstein was restricted to two essential requirements. First, the field equations themselves must obey the principle of relativity – they must have the same form in any frame of reference that moves in an arbitrary fashion, relative to any other, from the view of any one of these coordinate frames. Such an equation is called 'generally covariant'.

The second requirement was that his field equations were to represent a unique relationship between matter, on the one hand, and the space–time language that is to serve the role of facilitating an expression of the laws of this matter, on the other. Thus, Einstein set up an equation in which the appropriate variables, in representing matter, appear on the right-hand side, and corresponding field variables depending on the geometrical field, $g^{\mu\nu}$, and its variations in space and time, appear on the left. Einstein took the matter field on the right-hand side of this equation to be the energy–momentum tensor for the material system analysed, whether it would be our solar system or the whole universe!

Setting up a particular form for the field equations that would satisfy these criteria, Einstein then proceeded to look for physically connected solutions. If such solutions could lead to trajectories of matter that would duplicate at least some of the known data on the phenomenon of gravity, then this would be a first check on the validity of his theory, as applying to the phenomenon of gravity.

One of the first properties of matter that Einstein discovered from his equations was that, in the limit, when spatial separations between interacting matter become small enough, and when their relative speeds are small compared with the speed of light, then the ten equations in the components of $g^{\mu\nu}$, i.e. Einstein's field equations, break up into one important equation (in g^{00}) and nine less significant equations. If the ratio v/c may be set equal to zero, as an approximation, and the mutual separations between interacting matter is taken small compared with solar distances, then if v represents the speed of a planet relative to the Sun's position, or the speed of the Moon relative to Earth's position, etc., then the predictions of the equation in g^{00} becomes identical with the predictions of Newton's equations for universal gravitation. In this way, Einstein discovered that all of the numerically successful conclusions of Newton's theory of universal gravitation were also contained in the predictions of his theory of general relativity. But this was not enough to unseat a theory of gravity that held reign for the preceding 300 years! It was also necessary to show that Einstein's theory predicted new empirical facts that were not predicted by the earlier theory, if general relativity was to replace the classical theory.

In Einstein's theory it is the geometrical field, represented by the metric tensor, $g^{\mu\nu}$, that plays the role of the potential energy term in Newton's theory. The gravitational force in Newton's theory, in turn, identifies in Einstein's theory with certain rates of change of the ten components of the metric tensor, that in turn relate directly to the curvature of space–time. An important difference with the classical theory is that the terms that play the roles of potential and force in Einstein's theory are functions of both space and time, thus predicting that the gravitational force propagates at a finite speed, rather than just happening spontaneously, as in Newton's 'action-at-a-distance' view in classical physics. In this way, Einstein provided an explanation for the apparent action-at-a-distance between interacting masses that had so troubled Newton, according to his original writings on this subject, as an approximation.

Thus, it was Einstein's discovery that the (generally variable) curvature of space–time represents the corresponding variable force field of matter, which is an essential relationship in his explanation for the phenomenon of gravity.

With this theory applied to the phenomenon of gravity, a test body, such as the Earth, is predicted to move 'freely' along a special sort of curved path, if it would be in the vicinity of a body, such as the Sun, when the latter is represented by a curved space–time with a particular detailed form. The latter curvature, in this theory, is a representation for the existence of matter, as we have indicated previously. The slight departures from the cyclic motion that we know about, for example in the small changes, each year, in the time taken to for a planet to complete its yearly cycle, are primarily due to the classical gravitational forces exerted on this planet by the other planets of our solar system (as well as all of the other matter of the universe, in principle). A breakdown of the cyclic motion of a given planet of a solar system follows from Newton's theory itself, when its interactions with the other planets of the solar system are taken into account. However, there is observed a small amount of aperiodic motion of a planet that is not at all predicted by the classical theory. This will be discussed later on in this chapter, as one of the critical tests of Einstein's theory, when it was applied to the orbit of Mercury.

According to Einstein's approach, the predominant aperiodic effects, that are explained classically from the gravitational interaction with the other planets, are, here, due to departures from the curvature of space–time that result from taking account of the existence of the other planets of our solar system, as well as the Sun. Similarly, the motion of the Moon about the Earth can be described, within general relativity, in terms of the curvature of the space–time at the site of the Moon (the 'test body', in this case) and so on. In the latter case, such curvature of space–time would be a geometrical representation of the existence of the Earth.

While all of the successful numerical results of Newton's theory of universal gravitation are incorporated in Einstein's theory of general relativity, it should be understood that this incorporation is only in the sense that Newton's formal theoretical construction serves as a mathematical approximation for Einstein's theoretical construction, when certain terms are neglected in the latter theory. Still, these terms are there all the time! Thus, Einstein's theory does not contain Newton's theory – the latter only serves as a mathematical approximation for Einstein's equations, for computational purposes!

These two theories of gravity are entirely different – both from the conceptual side and the mathematical language that represents the respective concepts.

1 Einstein's theory entails a ten-component metric tensor gravitational potential, while Newton's theory entails a single-component scalar potential.
2 Einstein's explanation of gravity is in terms of a continuum field theory in which the gravitational force propagates at a finite speed (its maximum being the speed of light in a vacuum), while Newton's explanation is in terms of spontaneous 'action-at-a-distance', that is, the gravitational force is exerted by a body on another over vast distances, with no passage of time.
3 Einstein's metric tensor field, that underlies gravity, depends on four independent parameters – the space–time coordinates – while Newton's gravitational potential depends on only one independent parameter – the distance between interacting bodies.

Even at the relatively small separations, when the formal expression of Einstein's theory may be approximated by that of Newton, there are still differences in the respective theories, because the ratio of the relative speeds of the interacting matter to the speed of light, v/c, is not exactly zero, and because the associated curvature of space–time is not exactly zero. These differences were dramatically verified in experiments – thereby instigating the revolutionary change that science introduced to the 20th century, in regard to an explanation for the phenomenon of gravity, as well as our concept of space and time. This success also hinted at the need to pursue Einstein's theory even further, to see if a more general geometry might incorporate other types of forces, such as the electromagnetic and the nuclear forces.

The three extra predictions that came from the equations of Einstein's theory of general relativity were:

1 the small extra contribution to the aperiodic motion of a planet, as we have discussed above;
2 the bending of a light beam when it should propagate in the vicinity of a massive body; and
3 the *gravitational red shift* is the prediction that the frequency of radiation, measured in one region of gravitational potential, would be shifted towards a smaller frequency (longer wavelength) if this radiation is measured as it propagates into a region of less gravitational potential. For example, yellow light in the visible spectrum, as measured in the region of the sun, would be shifted toward the red end of the visible spectrum (longer wavelength) if measured in the vicinity of the earth.[1]

Let us now briefly discuss these effects in more detail.

The aperiodic motion in the orbit of Mercury (the smallest planet of our solar system) was first observed in the middle of the 19th century. This is the effect whereby it takes extra time, each year of the planet's orbit, in order for it to return to an equivalent place, relative to the location of the Sun. As we have discussed earlier, most of this aperiodic motion is due to the coupling of the other planets of our solar system to Mercury, in terms of the ordinary newtonian force of gravity, thereby perturbing its (otherwise periodic) motion that would be caused by the Sun alone. The contribution of the sister planets to Mercury's aperiodic motion can be calculated with great accuracy – due to our knowledge of their locations at all times, relative to that of Mercury. However, the magnitude of the aperiodicity was measured in the 19th century (by U. Leverrier[2]) and found to be a small amount greater than the classical magnitude of this effect due to the other planets. Thus there was a genuine discrepancy between the classical prediction about the aperiodicity of Mercury's orbit and the experimental fact. [In technical terms, this effect is seen as a precession of the perehelion of Mercury's orbit – corresponding to a rotation of the elliptical axes of the orbit, with the Sun at one of its focii.]

In that time, some astronomers attributed the discrepancy to the existence of another, yet unseen, planet. However, such a planet was not discovered, and the anomaly remained a mystery – until Einstein's field theory predicted this difference with great accuracy, in very close agreement with the experimental fact.

This was a very important supporting result for Einstein's theory of general relativity. Unfortunately, many physicists and astronomers did not take it too seriously – simply because the experimental result came earlier than the theory! [It seems to me that this is a fallacious reason for rejecting a theoretical explanation for a fact of nature – it should make no difference whether the experimental result comes before the theory that explained it or vice versa!]

The reason for this effect in general relativity is that the time parameter is not generally separable from the space parameters, implying that there would not be stationary orbits in the predictions of the theory – resulting, for example, in the aperiodic orbit of a planet. [Mathematically, this unification of space–time means that the solutions of the generally covariant equations, being functions of the four space and time parameters, are not factorizable into a product of a spatially-dependent function and a time-dependent function. The latter factorization is a prerequisite for the appearance of a stationary state in the solutions of the equations of the theory, resulting in the prediction of periodic motion.]

The second prediction that was a crucial test of general relativity theory – the bending of a beam of starlight as it glances off of the rim of the Sun – was a more impressive test to the physics community than the aperiodic motion of Mercury's orbit because the prediction was made before the experimental confirmation. During a total solar eclipse, the light from a distant star was traced out as it propagates past the rim of the Sun. Knowing where the star was supposed to be, it was seen to be somewhere else, thus confirming the bending effect of the Sun on the starlight – by an amount that was in quantitative agreement with the prediction of Einstein's field equations. One of the first groups of scientists to observe this effect was the Cambridge University expedition, headed by A. S. Eddington, around 1919.

The reason for this effect in general relativity is that the path of the light is that of a geodesic. The latter, in turn, is a representation of the existence of a massive body. Most of the path of the starlight between the emitting star and the observer on Earth is in (practically) empty space, where the geodesic is a near 'straight line', as in euclidean geometry. However, when the starlight approaches the vicinity of the Sun, the geodesics continuously approach the curve that represents this body, as we have discussed previously. In regard to the conceptual meaning in general relativity, one should not look on this effect as the motion of a photon of light with inertial mass, that is attracted to the Sun gravitationally, when it comes near to it. One reason for not accepting the latter classical explanation for the observation is that it does not predict the numerical result of this observation correctly, whereas the geodesic explanation of general relativity theory does make the correct quantitative prediction. Also, it is fairly well established that the photons of light do not have any inertial mass – the universality of the speed of light implies that it cannot be speeded up or slowed down by any external force, thus implying that the photon of light cannot have any inertial mass.

A third crucial test of general relativity, the *gravitational red shift*, was not confirmed experimentally until many years after the prediction was made.

In the 1950s, an experimental group at Harvard University, headed by R. V. Pound, used the recently discovered 'Mössbauer effect' to detect a gravitational frequency shift. They did this by comparing the frequency of gamma ray emission from a sample on the roof of their building with the frequency of emission of the same type of sample in the basement of the building. If the building was h cm high, then the gravitational potential difference is gh (where g is the acceleration due to gravity). Due to the extreme amount of resolution achieved with the Mössbauer experimentation, their measurement was accurate to one part in a million-million! Their observed gravitational shift was found to be in very close agreement with Einstein's prediction from his field equations in general relativity.

The reason for the gravitational red shift is that the time measure is determined by the curvature of space–time where it is made. Since the amount of space–time curvature in the vicinity of a mass, say a low point on the Earth, is greater than a point on Earth with a greater distance from its centre, and the latter two cases correspond to greater and less gravitational potential, respectively, there is a corresponding difference at these two points in regard to the time measures at these points. Since the frequency of emission of radiation – a measure of a number of cycles per second – is equivalent to a measure of time, it is anticipated that the frequency should shift, when it is measured in one place of potential compared to another place of a different potential. Again, the reader should be cautioned that it is not the classical newtonian potential energy of the Earth that, in this theory, is the cause of the shift of frequency. It is, rather, the different amounts of curvature of space–time, expressing different geodesics, that is the explanation provided by general relativity theory, even though, for convenience, we correlate the places of different mass effects, and therefore different amounts of curvature, with measurements of different newtonian potentials. That is to say, as with the previous two crucial tests, this is a consequence of the theory of general relativity; it is not a consequence of the classical theory. The positive results of general relativity are not even predicted qualitatively by Newton's theory of universal gravitation. It was a striking success for general relativity theory, in superseding Newton's classical theory of universal gravitation.

At this stage, then, we see that Einstein's theory of general relativity predicted all of the successful results of Newton's theory of universal gravitation, *and* successfully predicted three new effects that are not at all implied by the classical theory. His theory then met the criteria to unseat a theory of gravitation that held reign for 300 years. It also gave science a new perspective on the role of continuity and space and time, in the laws of nature and it pointed to new directions in order to comprehend the universe – from the domain of elementary particle physics to that of cosmology.

It is interesting to conclude this chapter with a speech that was given, by the playwright, George Bernard Shaw, in London in 1930, in honour of Einstein. The following [3] are excerpts from this speech:

> . . . a young professor rises calmly in the middle of Europe and says to our astronomers: 'Gentlemen, if you will observe the next eclipse of the sun carefully, you will be able to explain what is wrong with the perehelion of Mercury.' [This refers to the crucial test of general relativity having to do with the aperiodic motion of Mercury's orbit, as we have discussed above]. The civilized newtonian world replies that, if the dreadful thing is true, if the eclipse makes good the blasphemy, the next thing the young professor will do is question the existence of gravitation. The young professor smiles and says that gravitation is a very useful hypothesis and gives fairly close results in most cases, but that personally he can do without it. He is asked to explain how, if there is no gravitation, the heavenly bodies do not move in straight lines and run clear out of the universe. He replies that no explanation is needed because the universe is not rectilinear; it is curvilinear. The newtonian universe

thereupon drops dead and is supplanted by an einsteinian universe. Einstein has not challenged the facts of science but the axioms of science, and science has surrendered to the challenge.

In London great men are six-a-penny and are a very mixed lot. When we drink to their health and make speeches about them, we have to be guilty of scandalous suppressions and disgraceful hypocrisies. Suppose I had to rise to propose a toast to Napoleon. The one thing I should not possibly be able to say would be perhaps the most important – that it would have been better for the human race if he had never been born. Tonight, at least, we have no need to be guilty of suppression. There are great men who are great among small men. There are great men who are great among great men, and that is the sort of man we are honouring tonight. Napoleon and the other great men of his type were makers of Empire. But there is an order of man who gets beyond that. They are makers of universes and as makers of universes their hands are unstained by the blood of any human being.

NOTES

1 For a detailed derivation of these phenomena, see, for example, C. Møller, *The Theory of Relativity* (Oxford, 1967), Ch. 12.
2 See: R. J. Adler, M. J. Bazin and M. Schiffer, *Introduction to General Relativity* (McGraw-Hill, 1975), 2nd edn, Sec. 6.3.
3 B. Patch, *Thirty Years with G.B.S.*, (V. Gollancz, London, 1951), pp. 193–4.

20

Faraday's unified field concept

Strongly influenced by the 18th century thinker, Roger Boscovich, Michael Faraday, in the 19th century, introduced the field concept to physics with its application to the description and understanding of the phenomena of electricity and magnetism. He speculated that, contrary to the atomistic view of matter, its fundamental description should instead be in terms of continuous fields of force. These were to represent the strength with which matter affects matter, at the continuum of points in space and time. Calling this description 'fundamental' meant that it was the continuous field of force that was to be taken as the essence of matter, rather than considering this as a secondary (derivative) feature of a quantity of matter. The 'thing' feature of the atomistic view, for example in terms of its position, momentum and energy, all localized in space, was then taken to be secondary – to be derived later, as a consequence of the primary field of force – rather than the other way around.

To exemplify this conceptual difference between Faraday's continuum view and Newton's atomistic view, consider the interpretation of the (empirically confirmed) law of universal gravitation. If any two quantities of matter, with inertial masses equal to m_1 and m_2, should be separated in space by the distance R_{12}, then the force of attraction between them is

$$F = G\frac{m_1 m_2}{R_{12}^2}$$

Faraday interpreted this formula in a non-atomistic way as follows. Consider a potential field of force that could act on a test mass somewhere, if it should be there, as the continuous field,

$$P_2 = G\frac{m_2}{R^2}$$

where R is a continuously varying spatial point, measured from some appropriate spatial origin, for the problem at hand. The field of force, P, is then defined continuously everywhere, except at the origin, $R = 0$. If we now consider a test mass, m_1, to be placed at $R = R_{12}$, then the observed force at this place would be the coupling of this test mass to the field of force P_2 at this point of observation. That is, the force on the test body at the location $R = R_{12}$ is the coupling of this mass to the field:

$$m_1 P_2 = m_1 \frac{Gm_2}{R_{12}^2}$$

in agreement with the empirical expression for the gravitational force. The constant G is called 'Newton's gravitational constant' – it is the same for all gravitational interactions.

Faraday then asserted that the continuum of lines of force, defined by the field P for gravitation (or by other fields for other types of force, such as the cause of the magnetic alignment of a compass needle or the cause for the electrical repulsion between two bits of amber that had been rubbed with fur), are the fundamental stuff from which any theory of matter should be built. Faraday (and Oersted before him) did not believe that there are fundamentally different types of forces between matter and matter. He felt, rather, that all forces must be unified into some sort of universal force – an entity that would only appear as disconnected types of force under different sorts of experimental circumstances.

In Denmark, Hans Christian Oersted first discovered that a compass needle will become aligned in a particular way in a plane that is perpendicular to the direction of flow of an electric current. Faraday then interpreted the magnetic field of force (in this case, acting on the compass needle) as no more than a representation of an electric field of force in motion, that is to say, an electric force as observed in a moving frame of reference relative to a stationary observer (that of the compass needle). Since 'motion', *per se*, is a purely subjective entity, the implication then followed that the electric and the magnetic fields of force are actually no more than particular manifestations of a unified electromagnetic field of force. It also implied, because of the relativity of motion, that an electric field is nothing more than a magnetic field in motion! [This idea led to the invention of the dynamo – so important to the industrial revolution of the 19th century.[1]]

As we have discussed earlier, James Clerk Maxwell, in expressing Faraday's unified field theory in a rigorous mathematical form, showed further that such unification also incorporates all of the known optical phenomena – the ray properties of light (e.g. the focusing of light through lenses, etc.) as well as the wave features – its polarization characteristic (being polarized in a plane that is transverse to its direction of propagation, (as originally discovered by Augustine Jean Fresnel), its properties in combination with other optical waves yielding interference and diffraction, etc. The Maxwell field theory also predicted other radiation phenomena, such as radio waves, X-rays and gamma rays.

With his discovery of the unification of electricity and magnetism, together with optics, Faraday believed that this was only a first step toward a fully unified field theory of force – one that would incorporate the gravitational force as well.

Ingenious as he was, Faraday was not able to devise any experimental scheme that would verify such a unification of electromagnetism with gravity. For example, he was not able to convert electric or magnetic currents into a gravitational current, or to demonstrate the existence of gravitational radiation that could fuse with electromagnetic radiation.

From the theoretical side, there were also some unanswered questions on the validity of such a unification. An important one concerns the reason why, on the one hand, electromagnetic forces can be either attractive or repulsive, while, on the other hand, gravitational forces have only been found to be attractive.

In the century that preceded that of Faraday, Boscovich[2] was concerned with this question. He attempted to answer it by describing gravitational forces in terms of a continuous effect that matter has on matter, as a function of their mutual separation, in such a way that the attractive character of the force changes smoothly into a repulsive force when sufficiently small mutual separations are reached. However, he did not entirely succeed in this theoretical task, nor did Faraday explain the disparity between the natures of the electromagnetic and the gravitational forces, nor did Einstein resolve the problem in the contemporary period. But all of these investigators did believe that the approach of a fully unified field theory that would incorporate the electromagnetic and the gravitational forces, along with any other type of force that may exist in nature, such as the nuclear (strong and weak) interactions – that were not known yet in the 18th and 19th centuries – was indeed in the direction toward a general theory of matter. The aim was to construct a unified field theory that would incorporate the electromagnetic, gravitational and any other type of force to be discovered in the future – in a way that the universal force field would manifest itself as different sorts of force under correspondingly different types of experimental conditions, such as the manifestation of the magnetic force when the viewer is in motion relative to the reference frame of an electrically charged body.

In the 20th century, the other types of force that revealed themselves were in the domain of atomic and nuclear physics. The nuclear (strong) force, that binds neutrons and protons in an atomic nucleus, is the order of a hundred thousand times stronger than the repulsive force between two protons due to their mutually repulsive electric force. In this domain, the electric force between the protons is the order of 10^{40} times stronger than a newtonian force of gravity between them. Also in the nuclear domain there are the 'weak interactions', which are responsible, for example, for beta decay of radioactive nuclei – the conversion of neutrons (or protons) in nuclei into protons (neutrons), electrons (positrons) and neutrinos. In this domain, the strength of the weak interaction is around a hundred times weaker than the electromagnetic interaction. Both the nuclear (strong) and the weak interaction effectively 'turn off' as the mutual separation exceeds the size of a nucleus, around 10^{-12} cm. On the other hand, the electromagnetic and the gravitational interactions have infinite range – that is, they never really 'turn off'. In contrast with the electromagnetic and gravitational forces, the nuclear force has a part that is attractive and a part that is repulsive. Which one dominates depends on the conditions imposed, while the other is masked. Different types of nuclear experimentation, mostly involving the scattering of nuclear particles from each other, are able to project out the repulsive and attractive components of the nuclear force, separately, in the analyses of the data.

According to Faraday's view, then, a fully unified field theory must incorporate all of the different types of force – the gravitational, electromagnetic, nuclear, weak, . . . into a unifield field of potential force, with each generic type of force manifesting itself with regard to the behaviour of a test body, under correspondingly different types of experimental conditions.

Finally, it is important to note once again that 20th century physics has revealed the fact that the inertial features of matter are intimately related to what seems to be an entirely different sort of continuous field – the field variable that has been interpreted as a 'probability amplitude' in quantum mechanics. A fully unified field theory would then also have to incorporate the successful formulae and equations of quantum mechanics in order to describe the atomic domain satisfactorily. The formalism of quantum mechanics that is used today may then appear as an approximation within a more general field description that would unify the force manifestations of matter (gravity, electromagnetism, nuclear, . . .) and the inertial manifestations of matter (the feature of resisting a change in its state of constant motion, due to imposed forces). That is to say, a fully unified field theory must incorporate the actions of bodies on each other (the forces) and their respective reactions (their inertia).

Question Is there any logical reason, that is, aside from aesthetics, that all of the laws of nature must be derivable from a single, general theory?

Reply It is true, of course, that 'beauty is in the eye of the beholder', but in addition to the reaction that I have (or that you may have) to the idea of a complete, unified scheme for all of the different types of physical phenomena, as being more beautiful than a patchwork of disconnected theories, there is also an important logical reason for striving toward a general theory. A patchwork scheme runs into difficulties on the question of logical consistency, especially at the 'seams' where the patches are supposed to be joined.

Consider, for simplicity, two physical theories, initially constructed to predict two seemingly mutually exclusive sets of phenomena. Suppose that these theories are such that a part of the axiomatic basis of one of them is logically incompatible with some of the underlying axioms of the other. For example, one theory might assert, axiomatically, that one quantity of matter exerts a force on another only when they are in contact. The second theory may assert, in contrast, that forces between distant quanties of matter are exerted spontaneously, over any distance. Now it may happen that as the methods of experimentation improve, the conditions which dictate the criteria for using one of these theories or the other will fuse into each other – then requiring the use of both theories at once!

A well-known example is the theory of general relativity, replacing Newton's theory of universal gravitation. These are certainly logically dichotomous theories; if taken together they would be a logically inconsistent, single theory. Einstein's theory is based on the continuous field concept and the prediction of a finite speed of propagation of interactions between the material components of the system; Newton's theory is based on the discrete atomistic model of matter, and the notion of spontaneous 'action-at-a-distance'. If one should allow the speed of one gravitationally interacting body to be continuously slowed down, relative to another, and their separation is allowed to become sufficiently small, so that in a minute portion of space the curved space–time in their domain may be replaced with a flat space–time (the tangent plane to the point of observation in a curved space), then, Einstein's field equations can be approximated by Newton's equation for the gravitational potential. Trying simply to adjoin Einstein's theory to that of Newton then leads to the question of logical consistency when trying to decide precisely where should one 'turn off' the concepts of Einstein's theory and 'turn on' those of Newton's theory. There is no unambiguous solution to this problem! It is then for the sake of logical consistency that one cannot consider both of these theories as correct, even in the proper domains where they 'work'. To maintain logical consistency – that is, to be able to say that particular empirical data can be 'explained' by one theory and not another theory that is incompatible with the first – it is necessary to foresake one of these theories for the other, even though the foresaken theory mathematically approximates the accepted theory, under those experimental conditions where the foresaken theory 'works'. It is, in fact, the latter comparison that gives us hints about the character of a more general theory – through the requirement that this formalism must approach the form of the earlier theory, in the limit where the latter theory was empirically successful. This is known in the history of science as a 'principle of correspondence'. It has been used throughout the different periods of what Kuhn[3] has called 'normal science', and their transition through a scientific revolution, to the next period of normal science. It is in this sense that continuity persists throughout scientific revolutions – even though many of the ideas of the superseded theories are abandoned.

Question Is there any comparison between the evolution of ideas in science and the evolution of the human society?

Reply I think that the continuity of ideas in science is analogous to a continuity that persists in certain features of the evolution of the human society, in proceeding from one generation to the next. Certainly all of the concepts adhered to by an older generation are not maintained in all succeeding generations! But there are notions, values and facts that do persist. That which remains had usually been sifted, refined, re-evaluated, and taken in a new perspective. I would then contend that without these concepts and attitudes that had been passed on from one generation to the next in this way, there could have been no progress at all in our comprehension of the world, nor could there have been a positive propagation of values of a civilized society. Little as this progress has been since the infancy of the human race, I believe that it is nevertheless nonzero, as exemplified in the laws of society that we do have (at least on the books!) to protect the rights of all people.

On the notion of continuity, I am often reminded of a conversation I was told about between a former colleague of mine and his nine year old daughter, Suzie. One morning, at breakfast, Suzie asked her Dad, 'Where was I before I was born?' He replied, 'You were nowhere before you were born! You didn't exist!' Without any hesitation, Suzie replied, 'I don't believe it. It can believe that you may not know where I was before I was born, but I must have been somewhere!'

Suzie's Dad then took a blank napkin from the table and a pen from his pocket. He showed her the blank napkin and asked, 'What do you see on this napkin?' Suzie answered, 'Nothing'. Then her Dad drew a circle on the napkin with his pen and asked: 'What do you see on the napkin now?' Suzie answered, 'A circle'. Convinced of his argument, Suzie's Dad then asked her, 'Where was the circle before I drew it here on the napkin?' Suzie replied, again without hesitation, 'It was in the pen!'

I suspect, and I related this to my colleague when he told me about the conversation, that, perhaps because of her beautiful lack of inhibitions, Suzie had a more subtle, profound and scientifically valid answer to her own question than her Dad did! For the next question would have been: 'Where was the circle before it got into the pen?' This would have led to Suzie's father's mind, leading in turn to his experience with the world, . . . and so on, *ad infinitum*, until the entire universe would have been used up! This final answer may then have viewed the world holistically, in terms of all of its seemingly different manifestations (people, planets, galaxies, . . .) from the approach of a unified field theory!

NOTES

1 R. A. R. Tricker, *The Contributions of Faraday and Maxwell to Electrical Science* (Pergamon, 1966).
2 R. Boscovich, *A Theory of Natural Philosophy* (MIT, 1966).
3 T. S. Kuhn, *The Structure of Scientific Revolutions* (Chicago, 1970), 2nd edn.

21

Einstein's unified field concept

During his later years, Einstein attempted to show that, at least the electromagnetic field of force could be unified with the gravitational field of force, in terms of a more general sort of geometry than that which he found earlier to represent accurately the gravitational force alone. For if the gravitational force could be derived as a feature of the motion of a freely moving body in a space–time governed by the relations of a riemannian geometry, and if, in principle, there is nothing exclusive about the gravitational force, as compared with any other type of force in nature, then a generalization of the geometry was a natural step toward further extension of the theory of general relativity, which Einstein considered to be its completion.

After all, in addition to their inertial properties, gravitating bodies do possess electric charge, so that when brought sufficiently close together their mutual electromagnetic interaction does 'turn on', dominating the gravitational force in this region. But even when it does dominate, the gravitational force is present – they are there together, though one of them is masked. That is, in the limiting process, it seems unreasonable that at some (unprecisely defined) juncture, the riemannian geometry would suddenly 'turn off' and the euclidean geometry would simultaneously 'turn on', along with the action of the electromagnetic forces. It seems more reasonable, at least from Einstein's point of view, that there should exist a generalization of the geometry of space–time that would, in principle, unify the gravitational and the electromagnetic fields, under all possible conditions – giving an apparent view of pure electromagnetic forces at sufficiently small separations and an apparent view of pure gravitational forces at sufficiently large separations – but both actually being present at all times, with one or the other dominating.

In addition to the role of geometry in Einstein's unified field theory (a feature not present in Faraday's unified field approach), there is also the following important conceptual change. In Faraday, the field of potential force represents matter – it is its essence. The test body is then brought into the picture to probe this field. In Einstein's field theory, on the other hand, especially when it incorporates the Mach principle, the test particle is not, in principle, separate from the field of force that it is supposed to probe. Rather, there is one closed system to consider, with its underlying continuous field that incorporates, in a unified way, all interacting components. The test body is then abstracted from the description of this closed system in an asymptotic limit, where the particular component of the whole system only appears as a weakly coupled 'part'. This abstraction is carried out by taking a special approximation for the full mathematical expression of the closed system.

The two views are conceptually inequivalent. Einstein's unified field theory approach is a generalization of Faraday's in the sense of predicting, in principle, more physical features than are predicted by Faraday's approach. One reason is that in Faraday's field theory the test body is ambiguously defined; for example, it may have an electric charge if it is to probe the strength of an electric field of potential force, but none other of its fundamental properties are involved in the basic theory. In Einstein's theory, though, the test body is fully and uniquely defined in all of its characteristics – following from the asymptotic features of the entire closed system that one starts with. Thus, while more of the features of the physical system are implicit in Einstein's theory, all of the predictions from Faraday's theory are also contained in the predictions of Einstein's theory. Thus, Einstein's approach to a unified field theory is a true generalization of Faraday's approach to a unified field theory.

A major conceptual change from Faraday's unified field concept to that of Einstein, regarding what has just been said about the different ways in which the test body is introduced, is the difference between the 'open' system, that characterizes Faraday's theory, and the 'closed' system, that characterizes Einstein's theory. For example, Faraday's field of force is a linear superposition of fields – a (vectorial) sum of the force fields for each of the component 'influencers' in the system, that could act on a test body. Here, the test body is just one more object – whose field is summarized in terms of a set of particle-like qualities associated with its localization – such as charge, mass, magnetic moment, etc. Thus, Faraday is still talking about many individual 'things', with his field theory, although they are most fundamentally characterized by their separate fields of influence – later to be added in order to determine the total effect on a test body.

Einstein's unified field theory, when it is expressed in a way that is compatible with its meaning, and when it incorporates the Mach principle, is to represent only one thing – this is the single, closed system, not composed of separate parts. This is the universe. It is a philosophy that implies holism. The apparent view of a system in terms of separable parts is, here, only illusory. It appears that one of the inseparable components of the closed system is indeed a 'thing' that is weakly coupled to the rest of the system – like the coupling of our planet, Earth, to its entire host galaxy, Milky Way – so much so that we could extrapolate to the limit where they are actually uncoupled things! But the latter extrapolation is, in principle, impossible, according to Einstein's conceptual stand. This is because the test body is still only a particular manifestation of the entire closed system, just as a ripple of a pond is only a particular manifestation of the entire pond, rather than being a thing-in-itself. One cannot remove a ripple from the pond and study it on its own, as an independent 'thing', with weight, size, and so on.

Nevertheless, from the mathematical point of view, one can, with accuracy, assume that the weakly coupled entity (like the ripple) is practically a disconnected entity. One may then derive its motion as the dynamics of an independent body, from the equations of motion that would describe such an independent thing, perturbed by the rest of its environment. It is still important, however, that one must view the latter mathematical description as no more than an approximation for an exact description of the entire closed system – a system that is in fact without independent parts. The latter more general mathematical structure is quite different than the structure of the equations that describe the approximated situation. These differences might show up in other predictions, even when the system looks like it has an independent part in it, though weakly coupled to the rest.

Question How does the philosophy that underlies Einstein's unified field concept compare with earlier views of closed systems, such as those of Spinoza and the monad theory of Leibniz?

Reply The philosophy that follows from Einstein's unified field concept, attempting to express all physical manifestations of matter in a single conceptual scheme, does indeed bear some resemblance to the Leibnizian world as fundamentally one, without separable parts – from my understanding of his philosophy. Plato also had the view that the universe is a single, existential entity, though he believed that its form is changeable.

Leibniz[1] viewed the world in terms of what he called 'monads' – entities that many philosophers have likened to atoms. But Leibniz' monads were not really separable things, as are the constituent elements of the real, material world, according to the atomists of ancient Greece, or the atomists of the contemporary period – such as Bohr and Heisenberg. Leibniz' monads are different from separable atoms – they are more like the ripples of a pond, or the notes sounded by a violin string – as particular modes of behaviour of a closed system. Leibniz referred to his assemblage of monads as an array of reflections of one world, I would then interpret his word 'reflection' to mean 'physical manifestation'. With this interpretation, Leibniz would not be taking the atomistic view at all; rather, he would be taking an approach more similar to that of Einstein, where it is assumed that there is only one, real, objective world, without actual parts, but rather characterized by an infinite spectrum of possible physical manifestations (monads).

It is interesting to extrapolate from this holistic view of the material world to a universe that includes humankind. Assuming, with Spinoza[2], that such an extension is logically necessary, one might then view the human beings' inquiry into the nature of the universe (including themselves!) as based on an approximation that our consciousness can get into, thereby reflecting on underlying abstract relations about the nature of the real world. We may then extrapolate from these relations toward a more complete understanding of the universe – that fundamental existent of which the human being is, in principle, an inseparable component. According to this view, the human being's attempt to comprehend the world is not a matter of our 'looking in', as impartial observers, tabulating its physical characteristics, according to the responses of our sense impressions.

It seems to me to be, rather, a matter of the human being's intellectual reflections, introspections and deductions about the basic nature of a single, abstract, underlying reality – that existent that is the universe, and from which the human being derives particulars to be correlated with the physical reactions of our senses. Still, we must keep in mind the idea that there are infinitely more particulars that fall within the scope of our reasoning power, but are not within the domain of direct responses to our sensing apparatuses. Nevertheless, the latter (unobservable) particulars can play the important role of the logically necessary ingredients of a theory, that leads, by logical implications, to bona fide tests of our comprehension that do entail directly observable particulars.

There are critics of such a unified approach who argue that, according to Gödel's theorem, in logic, it is illogical to start at the outset with a complete system, because the investigators themselves must have the freedom to decide how, where and when to explore along one direction or another[3], i.e. they have the freedom to exercise a decision-making process, with their own 'free wills'. But this view tacitly assumes that the human being is indeed a separable entity, apart from the rest of society and everything else in our environments! Should one accept this assumption as a fundamental axiom, then Gödel's theorem could apply to the real world, outside of the logical system to which he applied it (a theory of arithmetic). One would then not be able to assert that there could be 'complete knowledge' to talk about in the first place. On the other hand, Gödel's theorem is not an a priori truth of nature! It is only as true as the set of axioms on which it is based. I, for one, do not accept the axiomatic basis of this theorem as relating to nature, since I believe (with Spinoza) that we are truly one with all of nature, not separable parts, but rather as manifestations of the holistic system that is the universe. Thus, I believe that there is indeed 'complete knowledge' – a total underlying order. But I do not believe that we can ever reach this total knowledge, because it is infinite in extent. Still, I feel that it is our obligation, as scientists and philosophers, to pursue this objective knowledge of the universe, bit by bit, continually criticizing and rejecting what is scientifically invalid, and holding on to what is scientifically true.

According to Spinoza's view, in which there are no absolute lines of demarcation in the world between 'observer' and 'observed', and the leibnizian monad concept of the universe – an existent without actually separable parts – the tacit assumption that separates the human being from the rest of the universe must be rejected. That is, rather than the atomistic view, in which the universe is said to be composed of many coupled, though separable parts – some of which are the collection of independent consciousnesses with their own free wills – the proponents of the unified field theory must view the world, with Spinoza, as a fully deterministic existent, that may exhibit an infinite manifold of intrinsic manifestations; yet where free will (actual individuality and separability) is only an apparent (illusory) feature, that is no more than a particular approximation for the oneness of the universe, as is the ripple of the pond example, discussed above.

Such a view of the universe, as a truly closed system, not only serves the purpose of providing an important heuristic function in the metaphysical approach toward a general theory of nature. It also has mathematical and logical consequences that do not follow from the type of universe that is taken to be an open set of things – a sum of individual parts. Thus, the differences of these two metaphysical approaches – one in terms of a closed system and the other in terms of an open system – imply mutually exclusive features in the physical predictions that, in principle, could be tested further to verify one of these approaches or the other.

The philosophy of the unified field theory approach that we have been discussing bears a resemblance to the philosophies of many of the previous generations of scholars, even much earlier in the history of ideas than Spinoza and Leibniz. For example, some of Spinoza's views and his method of logical demonstration, can be traced to Moses Maimonides[4], who wrote his main philosophical treatise, *The Guide of the Perplexed*, in the 13th century. I believe that these scholars saw the oneness of the universe to include humankind. Should society ever be able to accept this view, it could lead to a fully rational approach to science, as well as a higher ethical behaviour of all human beings – for it is a philosophic approach that implies humanism and a oneness with all of nature.

NOTES

1 G. W. Leibniz, in: *The European Philosophers from Descartes to Nietzsche* (edited by M. C. Beardsley) (Modern library, 1960), p. 288.
2 B. Spinoza, in M. C. Beardsley, *ibid.*, p. 140.
3 K. R. Popper, Indeterminism is not enough, *Encounter* **40**, 20 (1973).
4 M. Maimonides, *The Guide of the Perplexed*, S. Pines, transl., (Chicago, 1963).

22

The night sky

The aim of physics and natural philosophy is to provide us with some comprehension of the whole, real, existing world, including our experiences with it – from the microscopic domain of elementary particle physics (where present-day distances probed are the order of 10^{-14} cm and less) to the astronomical domain of galaxies and clusters of galaxies, and beyond, to the domain of the universe as a whole – the domain of cosmology (where present-day observations entail distances of the order of 10^{28} cm and greater – about 42 orders of magnitude greater than the domain of investigation of the elementary particle physicist!)

The earliest investigations of our surroundings in terms of explicit laws of nature came from our observations of the night sky. Perhaps this is because it seemed then, in pre-historic times, to be the most awe-inspiring, and even threatening. Perhaps it did not occur to humankind in those ancient times that there is something to understand about ourselves and our immediate surroundings, and about our place in the universe.

Long after the pre-historic times, in the relatively recent past (about 2500 years ago), there was intense interest among the philosophers in ancient Greece and Asia, in the order of the objects in the night sky – the Sun and the Moon, the planets, the stars, the comets, and everything else in the night sky and the day sky. The geocentric model of the universe was generally believed for about 1500 years, from Plato to Copernicus, who, in the 15th century, speculated that, rather than the Earth, the Sun is at the centre of the universe, with the Earth and its sister planets, and other objects of the night sky, circulating about the Sun. This was indeed an 'Earth-shaking' claim for the human race to believe! For one thing, it offended egos and the grandiose view we had of ourselves at the centre of the universe. But there were also other objections that could be raised. For if the Earth were indeed orbiting about the Sun (rather than vice versa), then if it takes one year to complete one orbit about the Sun, 93 000 000 miles away from us, it is not hard to figure out that the Earth must be moving extremely swiftly relative to the Sun's position – the order of 100 000 miles per hour! Copernicus' contemporaries did not have these figures, but they could reason that the Earth's speed must be extremely fast. They must have then wondered if Copernicus is right, why we are able to stand on Earth without being thrown off! Perhaps it was in part for this reason that most of Copernicus' contemporaries felt that his assertion was ridiculous.

About 200 years after Copernicus, in the 17th century, Newton's law of universal gravitation explained the fact that things would not be thrown off the swiftly moving Earth. It was a generation before Newton, when Galileo discovered – with the help of the recently invented telescope – that Copernicus was right about the Earth not being at the centre of the universe. But he went beyond Copernicus in discovering that neither is the Sun at the centre of the universe! – that, indeed, there is no absolute centre of the universe to talk about! For Galileo discovered that the concept of motion is strictly subjective – it is just as correct to say, from an Earth-observer's reference frame, that the Sun moves about Earth, as it is to say, from the Sun's frame of reference, that the Earth moves about it – so long as the laws of motion are independent of the reference frame in which they are represented. This is Galileo's principle of relativity, which was an important precursor for Einstein's theory of relativity.

Copernicus discovered that the geocentric model of the universe, originally discussed by Aristotle, about 16 centuries earlier, could be refuted when account is taken of the slight motions of the stars during a given night – indicating that indeed the Earth does move! Galileo substantiated the Copernican view when he observed moons orbiting about Jupiter. He had incontestable evidence, then, that indeed there are objects in the night sky that do not orbit about an absolutely stationary Earth at the centre of the universe, and that the Earth does indeed move relative to the rest of the matter of the universe.

These studies led to the general impression that we human beings, on Earth, and even our solar system with mother Sun, are not that important in the universe, sitting on a speck of rock that happens to be orbiting about an average-sized star, Sun, that itself is not at any special location of the universe. The implication is that there must, undoubtedly, be similar physical situations in the rest of this vast universe, where there could be other living species, some at the same stage of evolution as us, some at much more primitive and others at much more advanced stages of evolution.

During the time of Galileo's studies, his contemporary, Johannes Kepler, was studying the motion of the planet Mars. He discovered that its orbit, relative to the centre of the Sun, is *not* circular, as previously believed by Galileo and all of the earlier astronomers and philosophers for thousands of years. Kepler discovered that Mars has an elliptical orbit, with the centre of the Sun at one of the elliptical focii. Kepler's belief in the existence of general laws of the night sky then convinced him that all planets must move relative to the Sun in such a way that their orbits are similarly elliptical. But it wasn't until Newton discovered his law of universal gravitation, along with his second law of motion, that a mathematical proof was offered for

Kepler's conjecture. Kepler had made the very important contribution to physics that there must exist general laws that underlie all of the special cases, and this idea was very influential in Newton's thinking. The idea of a general order of the universe was proposed by many preceding philosophers and scientists, such as Pythagoras, who lived 2000 years before Kepler. Nevertheless, Kepler did enforce his view by combining accurate observations of the night sky with his theoretical speculations on the laws of physics[1].

Beside the enormous importance of Kepler's and Galileo's discoveries in physics and astronomy, Galileo introduced the analytical method into scientific studies. He invented the 'thought experiment' – intuitively asserting axioms about the laws of nature for ideal situations, then arriving at logical conclusions about the nature of matter and its motion. He would then test his conclusions experimentally, to check the hypotheses originally posed. His analytical method depended on precise correlations between the experimental facts and the mathematical description (which, in Galileo's analysis, was based on geometrical concepts). Newton then invented calculus in order to describe variable motion in a precise way – though Newton himself did not use this mathematical tool in his own writings in physics. Newton saw Kepler and Galileo as his primary mentors, when he said:

> I would not have been able to see as far if I had not been standing on the shoulders of giants!

From the times of Galileo and Newton, progress in physics took great strides forward, especially in the attempts to understand the motions of matter in the laboratory, such as a block sliding down an inclined plane, in terms of a precise mathematical description – that could then be tested not only qualitatively but also quantitatively. In this way it could be discerned how well a hypothesis in physics is in agreement with the empirical data, in quantitative terms.

Scientists tried to understand astronomical observations with equal precision. But, unlike their experiments that entail laboratory dimensions, it is difficult to manipulate astronomical bodies! Thus, in the domain of astrophysics (the physics of the night sky) one could only speculate, make predictions from hypotheses and then check with the results of the (not too precise) observations.

An early type of astronomical observation was that there are non-radiating objects in the sky, such as the planets and their moons, which simply reflect the light from the radiating objects of the sky – the Sun and the other stars. It was learned that our mother star, the Sun, is an average sized star and a member of an ensemble of stars, called the Milky Way – which, in turn, is only one out of an indefinitely large number of other galaxies of the universe, and that the galaxies cluster in regions of the sky while many other regions are dark.

The physical characteristics of these objects that we can measure are their luminosities and their relative positions in the sky, compared with our own position. The 'luminosity' of a star is a relative measure of brightness, as compared with that of another radiating body. An example of a standard of luminosity may be that of a candle – the unit of brightness called a 'candle-power'.

In regard to the positions of the stars relative to us, the astronomers measure angles and they estimate distances. One way to measure distance is in terms of the intensity of the radiation emitted by the source. The astronomers also measure the spectra of the stars, which are characteristic of particular atomic species. Thus, the astronomers may identify the material constituency of the stars, and trace their evolution, from birth to death. For when a star radiates, it is burning up its fuel, and the remaining 'ashes' are of a different make-up, thus they have a different spectral emission than the initial star. That is, from a spectra and the equations of state of a radiating star, it is possible to obtain an idea of the evolution of a star, from its birth until its death.

It has also been observed that when the spectra of stars of the same age, but at different distances from us, are compared, these spectra are shifted toward the red end of the spectrum (longer wavelength) for stars in galaxies that are further from us than those of closer galaxies. As we will see next, this experimental result, which was first noticed by E. Hubble, in the 1920s, indicates that the universe as a whole is expanding. That is to say, each of the constituent galaxies of the universe is moving away from the other galaxies of the universe.

One way to determine the distance of a star from us (which the ancient Greeks tried to do, in vain!) is to compare the intensity of its radiation I with the intensity of radiation emitted from a standard source. Classically, the intensity of emitted radiation from a spherically symmetric star – which we assume is a hot gas in thermodynamic equilibrium with its emitted radiation, called *blackbody radiation* – at the site of the absorber (the observer, say on Earth), R cm from the source, depends on $1/R^2$. Thus, if I_0 is the intensity of the standard source and I is the intensity of the radiation of the star whose distance is sought, then the ratio of intensities is: $I/I_0 = (R_0/R)^2$, where R_0 is the distance from us to the standard source, say the Sun. The unknown distance of the given star is then:

$$R = R_0 \left(\frac{I_0}{I}\right)^{\frac{1}{2}} = \frac{k}{I^{\frac{1}{2}}}$$

where $k = R_0 I_0^{\frac{1}{2}}$ is determined by the standard source.

Another way to measure the distance to a distant star, or galaxy of stars, is to utilize Hubble's discovery of the expanding universe. What Hubble actually discovered was that in the shift in wavelength from the radiation emitted from one galaxy to us compared with the radiation emitted to us from a more distant galaxy, that is R light-years from the first one (a light year is the distance travelled by light in a vacuum for one Earth year – of the order of a million-million miles), then he found the relation, that is now called the 'Hubble law', having the following form:

$$c(\delta\lambda/\lambda) = HR$$

indicating that there is a linear relation between the Doppler shift of wavelength, $\delta\lambda/\lambda$, and the separation of the receding galaxies. So, the reason that we are convinced that the galaxies are in motion, receding from one another, is that the existence of this Doppler shift, as we have discussed earlier.

If an emitter moves away from an observer, at v cm/sec, and if f_0 is the frequency of this radiation from the reference frame of the emitting star, and if f is the frequency measured in the reference frame of the other galaxy that is in motion relative to the emitter, then the Doppler shift is the following, when the relative speed of the galaxies, v, is small compared with the speed of light, $f = f_0(1 - v/c)$, thus yielding the following Doppler formula in the non-relativistic approximation, as we have discussed earlier, as one of the tests of special relativity theory: $(f_0 - f) = vf_0/c$.

Since the wavelength of the radiation relates to the frequency according to the formula, $\lambda = c/f$, it follows that

$$z = \frac{f_0 - f}{f_0} = \frac{\delta\lambda}{\lambda} = \frac{v}{c}$$

where $\delta\lambda = \lambda - \lambda_0$.

Hubble's empirical observation was that the wavelength λ in the reference frame of the more distant galaxy is greater than the wavelength λ_0 in the less distant galaxy, for a given frequency of radiation, indicating a shift of this wavelength always toward the red end of the visible spectrum – this is the 'cosmological red shift'. This observation means that the galaxies are moving away from each other, thereby indicating an expansion of the universe. In contrast, a blue shift would indicate that the galaxies are moving toward each other – a contraction of the universe. In an 'oscillating universe', the Hubble result would then indicate that there would be a continual periodic change between red shift and blue shift, though we are now in the red shift phase of a cycle. This oscillating model, however, is speculative, since all that we can be sure of at the present is that the universe is now in an expanding phase.

According to the Hubble Law, then, $z = v/c = HR/c$, where H is a universal cosmological constant, called Hubble's constant. This law of the universe, then, indicates that the speed of a moving galaxy, relative to another one that is Rcm away from it, depends linearly on their mutual separation. That is to say, when the distance between the two galaxies triples, so does their relative speed triple, and so on.

It is important to take note that the Hubble law was derived by assuming that the speed of the galaxies, relative to each other, is small compared with the speed of light. For if v should approach the magnitude of c, then the approximated formula, that was used for the Doppler shift, would no longer be valid and the linear relation between v and R would break down.

In any case, Hubble's Law for the expanding universe, $v = HR$ is one method to determine the distances of galaxies in the universe from each other, where the relative speed v is determined from the Doppler shift in wavelength, as discussed above.

It is important to keep in mind the approximations that were used in arriving at these formulae. First, the dependence of the intensity of radiation from a star or galaxy, I, on the inverse square of the distance from it to an observer, $1/R^2$, is the classical result, which depends on the description of propagating radiation (considered as a 'blackbody') in a euclidean space–time. If such observations entail sufficiently small distances, like the width of a room, of even the distance from the Earth to the Moon, then this would not be a bad approximation. But when the distances are intergalactic, general relativity theory requires the use of a curved space–time, governed by riemannian geometry, as we have discussed earlier. The distances estimated by astronomers, when they use the classical formulas, may then be quite in error. For example, if one should watch a ship move away from the dock, disappearing over the horizon, then if one should assume that the Earth is flat, the estimate of the ship's speed, as a function of how long it takes the angle subtended by the ship at the observer's eye to disappear, would be much greater than if we assumed at the outset that the ship is moving over a curved surface. [Legend has it that Columbus decided that the Earth was round rather than flat, when, in his youth, he watched the sailing ships depart from the dock in Genoa, determining how long it took them to disappear from view, and comparing this with other estimates he had of their speeds.]

Another approximation that we used in coming to the Hubble Law is the one mentioned above, that of taking the approximation for the Doppler shift in relativity theory, where it is assumed that the speed of one reference frame, relative to another, v, is small compared with the speed of light c. Again, these are not good approximations for the extreme situations where the whole universe is involved in the observations, since it would then be necessary, in general, to use the formalism of the theory of general relativity, and not assume that the relative speeds are generally small compared with the speed of light.

In the next, concluding, chapter of this book, we will focus on the different models of the structure of the universe as a whole, – the subject of cosmology – and compare the implications of general relativity theory in cosmology with those of the classical views.

NOTES

1 The views of many of the early pioneers in the problem of astronomy, from the ancient times to the present, are presented in an edited volume of their original works: M. K. Munitz, (editor) *Theories of the Universe* (Free Press, 1957).

23

Cosmology

The title of this chapter comes from the two Greek words, 'Cosmos' – meaning the universe – and 'logos' – meaning reason. Thus, cosmology refers to the reason, or order, that underlies the entire universe. The aim of the part of astrophysics that deals with the subject of cosmology is then to gain some comprehension of the physical laws that underlie the behaviour of the universe, as a whole.

In Chapter 22, in discussing the Hubble law, we covered the class of present-day astronomical observations that would describe the universe as a whole – though the actual job of interpreting these data is indeed quite complicated. Whatever theory of the universe, whether it is Einstein's theory of general relativity or Newton's universal gravitation, applied to the universe, or some new innovation, it would be required to explain and predict these data successfully, as well as making new predictions of yet unobserved phenomena in the cosmological domain.

When the entire mass of the universe is involved, then according to the theory of general relativity, results interpreted with the formulations of the classical theories could be entirely fallacious, since the space and time of the universe as a whole is curved, globally, and the Hubble Law, as well as the intensity–distance relation, that we discussed previously, should be consistent with their formulations in terms of the riemannian geometrical system. In principle, then, one should start out to describe the universe in global terms.

Astrophysicists have studied a few different models of the universe, the most popular one today being the 'big bang model'[1], which assumes that at some initial time, all of the matter of the universe was so close together that there were no individual atoms or nuclei. According to George Gamow's model[2], all matter was initially a charge-independent neutron fluid. It was so dense that there was no empty space to talk about. [There are known stars that have densities that approach this magnitude – the 'neutron stars'.]

According to the 'single big bang model', all of the matter of the universe was in its maximum density state at the initial time, and maximally unstable.

At this 'initial time', the matter is then thought to have exploded, with freed neutrons flying off in all directions, with great speeds.

We now know that once a neutron is free, rather than being inside of nuclear matter, it is unstable. In a matter of minutes, it decays into a proton, an electron and a neutrino. The protons, so-produced just after the 'big bang', then collided with other neutrons to form deuterons and tritons (bound nuclei, each one with a single proton and one with one and the other with two neutrons). The deuteron and the triton then combine to yield a helium nucleus (two protons and two neutrons, all tightly bound together), a neutron and gamma radiation. The produced neutron from this reaction is then free to start the cycle all over again. The processes of capturing neutrons and protons then continued as the universe started to cool down, building up successively to species of the larger atomic nuclei, then, as the universe cooled down sufficiently, they captured electrons in orbits, to form the charge-neutral atoms, until all of the stable atoms the Periodic Chart were formed. [Gamow says that all stable atomic elements of the universe were created in this way in the time it would take to fry an egg!]

Eventually, then, these atoms formed into molecules and then to form stars and galaxies of stars and then to clusters of galaxies, as we now perceive in the universe. The gamma radiation that was initially produced with the 'big bang' is called the 'primeval fire ball'[3]. It is assumed that it was produced homogeneously and isotropically. In recent years, there has been observed in the heavens an isotropic distribution of radiation that has been identified with the 'primeval fire ball'. One reason for believing this is that if the background radiation would be emanating from our galaxy, it would not be isotropic, since the distribution of stars in this galaxy itself is not isotropic. [At this stage the identification of this radiation background with the 'big bang' is controversial, because its distribution is not exactly isotropic, though it is almost so].

To describe this process of the 'big bang' with the theory of relativity, an interesting analysis was carried out by Robertson and Walker. They showed that, consistent with Einstein's tensor formulation of general relativity, but inserting an extra term into their equation – the 'cosmological' term – there is a special type of invariant interval of the riemannian space–time, with the form

$$s^2 = (ct)^2 - R(t)^2 L^2$$

where L is a dimensionless quantity and $R(t)$ is time dependent, with the dimension of length, and depending on the global time parameter, t, $R(t)$ is sometimes called the 'radius of the universe'. To arrive at this metric, which predicts the Hubble Law in a certain approximation, from the big bang model, Robertson and Walker[4] had to add several new requirements, in addition to relativity's requirement of the principle of relativity. One of these is the assumption that the matter of the universe is homogeneously and isotropically distributed. This is called the 'cosmological principle'. Another of their assumptions is that of a global time measure – a measure of time with respect to the absolute reference frame of the universe, as a whole. This would be the same time measure from all galactic frames of reference.

Question Doesn't this global time, measuring the expansion of the universe, contrast with the idea of time in relativity theory? As you have explained, isn't time supposed to be a subjective parameter that a particular observer may use to express the physical laws, in one reference frame or another, with different corresponding measures?

Reply Yes, I agree that there is a contrast here. But if the theory of relativity is indeed one of our abstract laws of nature, and a cosmological model, that disagrees with it, such as the Robertson–Walker model, turns out to be empirically correct, then this model could only be taken as a mathematical approximation for a more general description that is indeed covariant – i.e. where there would not be an absolute (global) time measure that is the same in all possible reference frames. This is in the same sense that Newton's 'action-at-a-distance' concept – implying that the attraction of the Sun to the Earth is only a function of their mutual separation – must be rejected when Einstein's field theory is accepted. This is in spite of the fact that Newton's equations are valid as a good mathematical approximation in the domain of experimentation where the predictions of the newtonian theory are in good agreement with the empirical data.

Question Were you saying that the 'big bang' model in cosmology is not compatible with the theory of general relativity?

Reply Yes, I believe that this is basically true. If the theory of relativity, and its revolutionary interpretation of space and time, are taken to be the starting point for a theory of cosmology, then the big bang model cannot represent a true theory that is compatible in an exact sense with this theory – because it is not covariant. That is, it entails an absolute time axis – called 'cosmic time' – whereby all time measures may be made with respect to an absolute origin of time, which is the beginning of the universe – the time when it all started! This is clearly incompatible with the concept of time as a subjective parameter, dependent on the reference frame from which it is measured.

Of course, it is opposite to the spirit of science to claim that any theory will ever show itself to be absolutely true! As scientists, we are supposed to probe predictions of a theory as far as we can. If contradictions and inconsistencies appear along the way, then the theory must be altered or, if necessary, abandoned. But so long as we are investigating the full set of implications of the theory of general relativity, it must be recognized that it does imply the necessity of a cosmological model that does not single out any special time axis, even though, from our particular platform in the universe, we observe that the galaxies of the night sky seem to be moving away from us, in the course of (our) time measure. Thus, I do not believe that general relativity applied to cosmology can, in principle, be compatible with the present-day cosmological model of the big bang.

Question Will the cosmology based on general relativity come from a better determination of the solutions of Einstein's tensor field equations?

Reply Not necessarily. Einstein's tensor equations are not the most general representation of his theory of general relativity. As we have discussed earlier, it can be shown that the symmetry aspects of general relativity require that there must be 16 field relationships at each space–time point, while Einstein's equations are only 10 field relations at each space–time point.

Einstein was fully aware that his tensor field equations were not the final, most general expression for his theory. In his *Autobiographical Notes*, he made the following comment about his tensor field expression for his theory[5].

> Not for a moment, of course, did I doubt that this formulation was merely a makeshift in order to give the principle of relativity a preliminary closed expression. For it was not anything more than a theory of the gravitational field, which was somewhat artificially isolated from a total field of as yet unknown structure.

I have found, in my own research programme in general relativity theory, that there is a more general representation of this theory than the tensor form[6], – a theory in which the basic variables have 16 components, rather than 10. The basic variables are four-vector fields, in which each of the four components are quaternions rather than real number variables. The 'quaternion' is a different sort of number, discovered by the great Irish physicist and mathematician in the 19th century, William Rowan Hamilton. It has four independent components, rather than a single component, as in the case of real numbers, but its most revolutionary aspect is that the multiplication of quaternions yields a product that depends on their order of multiplication – i.e. the quaternions are not commutative under multiplication. I have found that a generalized kind of variable for general relativity is this sort of 16-component field, and that the fundamental field equations are then 16 in number rather than 10. We have discussed earlier the implications of the quaternion field in regard to the problem of the 'twin paradox', in Chapter 12. The quaternion formulation also makes other sorts of new predictions in the problem of cosmology. For example, it predicts that galaxies must rotate about an axis of symmetry – a result that is in agreement with the empirical facts, and is not, in principle predicted by the tensor form of Einstein's theory. In the problem of cosmology, the quaternion form predicts the possibility that the present expansion of the universe is only a phase of a continual expansion-contraction oscillating universe, in which the galaxies move in spiral fashion relative to the rest of the universe, rather than isotropically, as is advocated by the current views of the big bang cosmology[7]. There are many more predictions of this generalization in cosmology, and generally in physics – particularly in the domain of elementary particle physics[8] – that will be explored in future studies of this research programme.

Question You seem to be more interested in applying the theory of general relativity to the problem of elementary particles than to astronomical problems. Why is this?

Reply I am equally interested in applying this theory to both of these domains of physics. But if I concentrate more on elementary particle physics it is because I feel that the problem of matter is of primary importance at the present stage of physics. The requirement of understanding the elementary nature of matter from first principles has been with us since the earliest theoretical investigations of cosmology and matter in ancient Greece. But it has been especially so in the contemporary period, since the 1920s, when attempts were first made to fuse the quantum and relatively theories. The main attempts in physics, in our time, have been based on fully adopting the conceptual basis of the quantum theory, at the expense of abandoning the conceptual bases of relativity theory. These attempts have thus far been unsuccessful.

The idea of my research programme is the one proposed originally by Einstein – that perhaps the resolution of the problem of matter lies in fully adopting the axiomatic basis of the theory of general relativity, even if this may be at the expense of abandoning the conceptual basis of the quantum theory. Since the quantum approach has not yet been successful in yielding at least a demonstrably mathematically consistent theory, that is compatible with the requirements of both the quantum and the relativity theories[9], the objective scientist must not deny the possibility that the resolution of the problem of matter lies in the direction of Einstein's theory of general relativity, when it is fully exploited – both in regard to its mathematical form and its conceptual foundations.

A second reason for my concentration on applications of general relativity theory in elementary particle physics is that the type of experimentation that is done here is much more varied and entails much better resolution than the experimentation in astrophysics. In principle, it should be possible to acsertain enough information about the general field for a closed system from a study of the elementary particle region that might allow conclusions to be made with regard to the astrophysical domain! This is because we are talking about one field in this theory – a set of connective relations, mapped in all of space and time. It is then anticipated that some information could come out of this research, that explores the elementary particle physics domain, that could tell us something about the physical domain of cosmology – as remote as this may sound!

Question   If, according to the 'big bang' model, all of the universe was initially bunched up in a tiny region, and then exploded outwards, doesn't this imply that there is an absolute space in which matter sits, and then explodes into?

Reply   According to the theory of general relativity, matter is not in space. It is rather that the mutual interactions of all matter of a closed system defines the space. In the initial stages of the 'big bang', space was highly curved, everywhere. It is important to understand the meaning of this statement in relativity theory. It is not meant to say that there is a thing, called 'space', into which we put matter, and that one of the physical features of this thing is that it is highly curved when matter is in it!

The statement about the curvature of space–time, as we have discussed earlier, is that it is a feature of the language that is to represent mutually interacting matter of high density. But space, *per se*, is not an existent, independent of matter – e.g. there is no 'outside' of matter to talk about. There is only a universe of highly dense matter. With this view, when the expansion starts, at the big bang explosion, the decrease of the curvature of space–time expresses this physical fact about the diminishing of the matter's density. Thus, this expansion process should not be interpreted, within general relativity theory, as matter exploding into empty space!

Similarly, in regard to time, the explosion of the big bang happened at a time, also referred to only with respect to features of the interacting matter. 'Time' is undefined apart from matter. 'Before' and 'after' are simply undefined terms, apart from the mutual relations the define the matter of the universe, whether this mass is dense or rarefied. With this approach, the 'where' or the 'when' do not refer to any aspect of empty space or absolute time, just as 'blue', or the sound of falling rain drops, have no relation to the concept of empty space!

The idea is that space and time, as well as the logic that relates these coordinates (the algebra and geometry of the space and time) is only a language, invented for the purpose of facilitating an expression of the laws of nature, which are laws of matter. This was quite a revolutionary idea that introduced the theory of relativity to 20th century physics.

Question   How might the relativistic cosmology, that you have been explaining, incorporate the 'cosmological principle', that the standard theories seem to evoke – the idea that (apart from local irregularities) the universe should appear the same, from any spatial point?

Reply   The cosmological principle is an assertion about the uniformity of the distribution of matter in space. The assertion of the principle of relativity is also about uniformity, but in a quite different respect. For here it is assumed that the laws of nature, rather than the spatial distributions of matter, must be independent of the reference frame from which they may be viewed and expressed. That is, according to the theory of relativity, the set of basic relations that underlie our comprehension of the universe is uniform, rather than the particular way that material objects arrange themselves in space.

I cannot help suspecting that if the natural laws that were responsible for the anisotropic and inhomogeneous 'local' distributions of the matter of the sky – the spiral shapes of the galaxies, the clustering of galaxies, the presence of large expanses of 'dark matter' in the universe, etc., are the same as the laws that are responsible for the distribution of matter throughout the entire universe, then there should be some similar non-uniformity in the matter distribution of the universe, globally! This would be in contrast with the assertion of the cosmological principle.

Question I have an historical question. Does Einstein's relativistic cosmology depend, in any way, on the earlier concepts involved in Newton's cosmology?

Reply To answer this question, it is necessary to start with Newton's laws of motion. His first two laws entail the effect on a single body at a time, as caused by the total force due to all other bodies of a composite of masses, generally, e.g. the effect of the rest of the universe on a single galaxy.

On the other hand, Newton's third law of motion – for every action there is an equal and oppositely directed reaction – entails at least two bodies at a time. What the third law involves, then, is a completion of the dynamics of a material system, by including the reaction of the source of a force, when it acts on a given body. In this view, then, the complete description of the material configuration under consideration is closed, because it takes account of all couplings within the entire system, e.g. the totality of all galaxies of the universe. On the other hand, Newton's first two laws of motion – the assertion that an external force causes the body to accelerate (or the lack of such a force keeps a body at a constant speed or at rest) – entails only one body at a time, without taking account of the reactions of the material sources of the forces when they act on the given body.

The latter view, in regard to Newton's first two laws of motion, describes an open system and it is particularistic (atomistic), while the closed system, implied by this third law of motion, is holistic, i.e. it is in principle without parts. That is to say, in the holistic view, one may not remove any of the components of the system without changing the physical characteristics of the entire system – changing it into something else. On the other hand, the open system is characterized by one (or more) components that may be removed without in any way altering the basic characteristics of the entire system.

The conceptual basis of Newton's first two laws of motion is incompatible with that of the theory of general relativity. But the basis of his third law of motion is indeed compatible with the basis of general relativity, in regard to the elementarity of the closed system. Thus, it is my view that Newton's third law of motion is one of the most important precursors for Einstein's theory of general relativity. This is especially important in regard to the problem of cosmology. [The other two important precursors for Einstein's general relativity, in my view, are Galileo's principle of relativity and the field concept of Faraday, as we have discussed previously.] It seems to me, then, that these three precursors for Einstein's general relativity, and its application to the problem of cosmology, are threads of truth that have persisted through the centuries to our own period in the history of science.

Question What are some of the empirical differences between Newton's and Einstein's cosmology?

Reply One feature of general relativity, in cosmology, that contrasts with Newton's physics, is that the terms that play the role of force in Einstein's theory are not positive-definite. The gravitational forces, in Newton's cosmology, are only attractive. The force being non-positive definite in relativity theory, on the other hand, means that under certain physical circumstances, that are unique to the system considered, both attractive and repulsive forces are present, but one dominates the other. Thus, if the physical circumstances should change, a dominating attractive force (or a dominating repulsive force) could change to a dominating repulsive force (or to a dominating attractive force).

In the problem of the initial stages of the presently observed expansion of the universe, there was a time, generally associated with the 'big bang', when the repulsive force in general relativity dominated the attractive force, when the density of matter and the relative speeds between the components of the system were sufficiently great, thereby causing the explosion to start, driving the matter components away from each other. As time goes on, the density of the matter of this closed system as well as the relative speeds, decrease to the point where the attractive forces will dominate the repulsive forces. The expansion phase of the universe will then change to a contraction phase, and the material of this closed system will implode on itself, continuously increasing its matter density – until the critical value where the repulsive forces once again dominate, and the contraction once again changes to an expansion. This is an 'oscillating universe cosmology'. It implies that the last 'big bang' was the beginning of the present cycle of the oscillating universe. But the theory says nothing about when it all started! The latter, in my view, is not a question about physics; it rather concerns theology. But the oscillating universe cosmology, that we arrive at by fully exploiting general relativity, does not even address this question: how did the matter of the universe get into the unstable state, from which it exploded, in the first place? Many of my colleagues have been answering that 'this is the moment when God created the universe', or 'one should not ask a question like this because there were no laws of nature before the 'big bang' happened. But, in my view, these are not valid answers to a bona fide question in physics – they are non-scientific answers for a question in science!

In my view, the oscillating cosmological model is the only present-day approach that is compatible with the theory of general relativity. And it is compatible with the observations that indeed the matter of the universe is mutually repelling, rather than attracting. Those who uphold the newtonian view, that only attractive forces describe gravitation, say that in general relativity the expansion is not a repulsion of matter (the galaxies of the universe), but it is rather an expansion of the space that contains this matter. This explanation is not compatible with the theory of relativity because, in its view, space is not a thing in itself that can do something, or not, that is physical! 'Space' is no more than a part of a language that is there to facilitate an expression of laws of matter. Thus, if we see that the galaxies are indeed moving apart from each other, in accordance with the empirically concluded Hubble Law, this is a consequence of physical forces between the interacting matter – a repulsive gravitational force and not a property of space itself!

Summing up, then, the force of gravity, according to Newton's theory of universal gravitation, is only attractive, while the gravitational force in Einstein's general relativity contains both attractive and repulsive components at all times, with one of these or the other dominating, depending on the physical conditions of the closed system that are imposed. The salient point here is that the experimental facts, i.e. the observed expansion of the universe, is compatible with the predictions of the theory of general relativity and not with Newton's theory of universal gravitation, in the problems of cosmology.

Question I have read in the writings of some cosmologists that while the relativity theory you have been discussing is true on the local level, it is not so on the global scale of the entire universe – because there is only one universe and therefore it must be in regard to an absolute frame of reference.

Reply The relativistic cosmology rejects the notion of an absolute reference frame of the universe – a notion believed by Newton as well as some of the modern-day cosmologists. In the approach of an absolute frame of the universe, it is claimed, as you said, that because there is only one possible universe, the space and time coordinate system that is there to represent it, as a whole, is similarly absolute. But this conclusion is clearly *non sequitur*. That is to say, just because the universe as a whole is absolute (by definition – since it is all that there is, therefore it cannot be relative to anything else), this does not logically imply that the space–time language that we use, from our particular frame of reference, to express its laws, is similarly absolute!

Analogously maintaining the absoluteness of the meaning of a sentence in verbal language, such as 'a brown cow eats green grass', does not require a universal verbal language to express this sentence! It may be expressed in various and assorted cultural/verbal reference frames – English, Chinese,

Swahili, Hebrew, . . . while still maintaining the single, invariant meaning of the sentence. This is indeed the idea of Einstein's principle of relativity, when it is applied to the laws of matter.

A language generally consists of a set of words and a logic that connects them, in order to give meaning to the sentences that are made up from them. Such a logic of ordinary verbal language is its syntax, such as the subject–predicate relation. The 'words' of the space–time language of relativity theory are the space and time coordinates that we correlate with spatial and temporal measures. The logical relations between these 'words' are in two parts – algebra and geometry.

The algebraic logic of space–time in relativity theory is in terms of an 'algebraic symmetry group' – called the 'Einstein group'. This is the set of continuous transformations in space and time that preserve the forms of the laws of nature, from one reference frame to any other. The implication of the underlying Einstein group is that the mathematical functions that solve the laws of nature are continuous functions of the space and time coordinates, everywhere! Thus the field variables that are the solutions of the laws of nature (laws of matter) can have no discrete cut-off in space, anywhere. This conclusion then implies that any atomistic model of matter must be rejected. Rather, the matter components of any material system, all the way up to the universe as a whole (cosmology), are the continuous modes of a holistic field, analogous to the ripples of a continuous pond.

We see, then, that the basis of the theory of general relativity, and relativistic cosmology in particular, entails a purely holistic, continuum view of the universe. The galaxies and their constituent stellar compoents are then essentially not more than the modes of a single continuum, rather than separable 'parts' of the universe.

Question Is it possible that in the first moments of the 'big bang' – the single big bang that everyone talks about today, or the beginning of the present cycle of an oscillating universe, as you described – the rules of quantum mechanics had something to do with the way the universe started?

Reply There is a current view in cosmology, called the 'inflationary model', that attempts to unify notions of the quantum theory with general relativity, wherein the energy–momentum tensor, that serves as the source term for the geometrical fields in Einstein's equations, is taken to be that of the ·physical vacuum energy – this is the set of virtual particle–antiparticle pairs and radiation, that constitute a vacuum, according to the present-day scheme called 'quantum field theory' – the relativistic extension of quantum mechanics. This 'vacuum energy' is taken to be the cause of the original expansion of the universe. The scenario is as follows[10].

Initially, that is, before there was any meaning for the word 'time', the universe was a 'quantum vacuum', entailing a geometrical structure composed of 10-dimensional 'strings'. This 'pre-matter' collection of strings is described abstractly in the four dimensions of space–time as well as six additional

dimensions that entail a unification of the four fundamental forces in nature
– the gravitational, electromagnetic, weak and strong (nuclear) forces. At
this initial stage of the universe, the interaction of the strings was 'non-
singular' since, unlike the point particles of ordinary matter (which did not
yet exist) the strings are not located at singular points; rather, they are
smeared out over a small space.

The perfect symmetry of the 10-dimensional string universe was then said
to be destroyed by random quantum fluctuations of this world. The 'string
universe' was then inflated from the domain of micromatter to that of
cosmic proportions, leading to the formations of galaxies and clusters of
galaxies, in due time. Thus, the initial symmetry of a homogeneous and
isotropic universe, when the 10-dimensional string structure was in place,
was broken to a lower symmetry, where the gravitational and electromagnetic
forces became long range (and themselves of different strengths) and the
weak and strong (nuclear) forces became short range, of nuclear dimensions.

The strings then dissolved into gravitational radiation, in the forms of
'gravitons' – the quanta of Einstein's gravitational metrical tensor field.
Thus, the 10-dimensional space of strings reduced to the ordinary space–time
of elementary matter that we now have in fundamental physics. Such were
the changes that were supposed to precede the initial 'big bang', occurring
in the order of 10^{-35} seconds!

The physical processes that were involved in this critical 'inflation' of the
universe are said to be analogous to the phase change in the state of matter
that occurs at a critical temperature (and/or pressure), such as the change
from ice to water – thereby expending the latent heat energy to transfer
one of these phases of matter into the other. In the cosmological problem,
according to the string model, there is an analogy for the latent heat in
playing the role of restructuring the universe, from the totally symmetric
system of strings to the less symmetric system of ordinary matter, that has
emerged from the big bang, to the present state of the universe. Thus the
criticality of the phase change from ice to water is taken to be analogous
to the criticality of the phase change of the universe, from its more
symmetric string structure to its less symmetric phase of ordinary matter.
The symmetry change in the case of ice becoming water is seen in the
symmetric crystal structure of ice decreasing to the symmetry of the
amorphous liquid state of matter.

As I understand the scenario from the initiation of the big bang onward
in time, the assumption is made that the gravitational force between ordinary
matter can only be attractive, as in Newton's theory of universal gravitation
(but not like Einstein's theory of gravitation, as we have just discussed). It
is then postulated that, initially, there was only the physical vacuum, and
out of this there appeared a new sort of matter, coming with the breaking
of the string universe symmetry. This new matter is different than ordinary
matter in that it repels ordinary matter, rather than attracting it. Also, it
is quite unstable. This is the 'Higgs field' of elementary particle physics –

a type of matter that is essential to the success of the present-day theory of elementary particles, called 'quantum chromodynamics' – the so-called 'generalized gauge theory'.

With the latter model, then, in the first few microseconds of the universe, the Higgs field proceded to decay into ordinary matter that, in turn, was repelled by the rest of the Higgs field, until the entire Higgs field of the universe was used up, leaving the ordinary matter to continue in its expansion. This is the scenario that is meant to explain the presently observed expansion of the universe.

Question Do you have any criticism of this 'inflationary universe cosmology'? It seems to me to be rather *ad hoc*!

Reply While the inflationary universe cosmology is an interesting speculation about the first microseconds of the universe, attempting to explain its formation and expansion, I would like to make four major points of criticism. First, there is no empirically or mathematically conclusive evidence for this cosmological model, nor for the existence of the 10-dimensional strings. Secondly, there is no evidence for the existence of the Higgs field – though a great deal of work has been done in the past decades of high energy physics experimentation to find it.

Thirdly, when the theory of general relativity alone is fully exploited, there is no need for a new matter field in order to explain the initial mutual repulsion of the matter of the universe, to start off the presently observed expansion. As we have discussed earlier, the fields that play the role of force in Einstein's field theory are not positive–definite. Thus, contrary to Newton's theory of gravity, where the gravitational forces can only be attractive, Einstein's general relativity already predicts, in principle, both attractive and repulsive gravitational forces. The theory then entails the possibility of explaining the dynamics of the universe in terms of gravity having both the repulsive and the attractive component, with one or the other dominating under differing physical conditions of the universe – thus explaining the latest 'big bang' in dynamical terms, i.e. from the point of view of physics!

The fourth point of criticism is that there is no evidence that the theory of general relativity is logically or mathematically compatible with the quantum theory. That is to say, the inflationary model assumes at the outset that there exists a quantum theory of gravity. Not only has this possibility never been established, but there are bona fide technical reasons to believe that such a unification is impossible to achieve, if it depends on a union of Einstein's theory of general relativity and the quantum theory.

If the quantum theory and the theory of general relativity are indeed incompatible in this way, then one of these theories or the other must be abandoned, while properly generalizing the maintained theory so as to show conclusive predictions about the physics of the universe as a whole. If the quantum theory is the one that is to be maintained, then with the inflationary model, we would have to say that the universe was created by a 'quantum fluctuation' – an intrinsically probabilistic event! If the theory of general relativity gives the correct explanation, then there is no scenario about the 'initial' creation of the universe. But whatever is said about the universe is not framed in terms of probabilities – there are only predetermined events in this approach, including the model in which the universe is cyclic in its expansion and contraction – just as the seasons on Earth are cyclic in this way. Perhaps it is true that what we see locally reflects the global situation! In this case, we must conclude that the human race is not at the centre of the universe. But this should not discourage us! We should be happy that we are integral components of this glorious universe, and that we are endowed with the gift of rational thinking – to be able to comprehend it, at least in part.

NOTES

1 J. Silk, *The Big Bang* (Freeman, 1989), revised, updated edition.
2 G. Gamow, On relativistic cosmology, *Reviews of Modern Physics* **21**, 367 (1949).
3 A. Penzias, Cosmology in microwave astronomy, in: *Cosmology, Fusion and Other Processes* (edited by F. Reines) (Colorado, 1972).
4 A derivation of the Robertson–Walker metric is given in: R. Adler, M. Bazin and M. Schiffer, *Introduction to General Relativity* (McGraw-Hill, 1975), 2nd edn, Ch. 12, 13.
5 P. A. Schilpp, (editor) *Albert Einstein – Philosopher – Scientist* (Open Court, 1949).
6 M. Sachs, *General Relativity and Matter* (Reidel, 1982), Chapter 6.
7 M. Sachs, Considerations of an oscillating spiral universe cosmology, *Annales de la Fondation L. de Broglie* **14**, 361 (1989).
8 Several of these examples from the elementary particle physics are demonstrated in: M. Sachs, *Quantum Mechanics from General Relativity* (Reidel, 1986), Ch. 9.
9 Some of these difficulties are discussed in detail in: P. A. M. Dirac, *The Principles of Quantum Mechanics* (Oxford, 1958), 4th edn, Sec. 81. Also see: R. P. Feynman and A. R. Hibbs, *Quantum Mechanics and the Path Integral* (McGraw-Hill, 1970), Ch. 9 and p. 260.
10 A review of the relation of cosmology to elementary particle physics, in the context of the quantum theory, is given in Silk, *ibid.* Further discussion is in: S. W. Hawking, *A Brief History of Time – From the Big Bang to Black Holes* (Bantam, 1988).

Index

Printed and bound by CPI Group (UK) Ltd, Croydon, CR0 4YY

01/11/2024

01782614-0017